DATE DUE

DEMCO 38-296

DESERTS
The Encroaching Wilderness

DESERTS

THE ENCROACHING WILDERNESS

A World Conservation Atlas

Edited by Tony Allan and Andrew Warren
Introduction by Mostafa Tolba

New York
OXFORD UNIVERSITY PRESS
1993

General Editors
Professor Tony Allan
Dr Andrew Warren

Consultant Editors
Per Rydén
Dr Adrian Wood

Contributors
Dr Clive Agnew
Professor Tony Allan
Dr Philip Denwood
Crispin Hawes
Dr Ian Livingstone
Dr Keith McLachlan
Alan Martin
Dr Peter Moore
Professor Harry Norris
Professor Cliff D. Ollier
Professor Richard Reeves
Liane Saunders
Andrew Smith
Dr David Thomas
Dr Bryan Turner
Dr José F. Araya Vergara
Dr Andrew Warren
Giles Wiggs
Jonathan Wild
Dr Adrian Wood

Senior Executive Editor
Robin Rees

Senior Editors
Alfred LeMaitre
William Hemsley

Editors
Stephen Luck
Gavin Sweet

Proofreading
Fred Gill

Indexing
Kathie Gill

Art Editor
Iona McGlashan

Senior Art Editor
John Grain

Art Director
Tim Foster

Cartographic Editor
Andrew Thompson

Picture Research
Kathy Lockley
Anna Smith

Production
Sarah Schuman

Maps
Lovell Johns Limited,
Whitney, England

Published in the United States of America by Oxford University
Press, Inc., 200 Madison Avenue, New York, NY 10016

Oxford is a registered trademark of Oxford University Press

Edited and designed by Mitchell Beazley International Limited,
Michelin House, 81 Fulham Road, London SW3 6RB

Copyright © Mitchell Beazley International Limited, 1993

Library of Congress Cataloging-in-Publication Data
Deserts: the encroaching wilderness: a world conservation atlas/
  edited by Tony Allan and Andrew Warren
    p. cm.
  Includes index
  ISBN 0–19–520941–9
  1. Deserts. I. Allan, Tony. II. Warren, Andrew.
  GB611.D48 1993
  508.315'4–DC20   92–31992
                CIP

ISBN 0–19–520941–9

Printing (last digit) 9 8 7 6 5 4 3 2 1

Although all reasonable care has been taken in the preparation of this
book, neither the Publishers nor the contributors or editors can accept
any liability for any consequences arising from the use thereof or from
any information contained herein.

Typeset in Plantin Medium by Servis Filmsetting Limited, Manchester
Reproduction by Mandarin Offset, Singapore
Produced by Mandarin Offset
Printed and bound in Hong Kong

Measurements Both metric and imperial measurements
are given throughout.

Billions The usage of the word billion varies between countries, but in
all cases here it refers to thousand millions (1,000 million = 1 billion).

Jacket pictures Background: Planet Earth Pictures/Hans Christian
Heap. Front: NHPA. Back: Hutchison Library/Dave Brinicombe.

CONTENTS

INTRODUCTION

Two images of deserts hold the popular imagination. Both are wrong, as this book handsomely shows.

The first is that deserts are lifeless expanses of sand, empty except for the occasional nomad. In fact, deserts are home to an extraordinary web of life, harbouring a host of plant and animal species. Despite harsh conditions, deserts are also some of the most exciting, varied and, to humans, useful ecosystems in the world. Some of our most important food crops, including several cereals, come originally from areas that scientists would define as deserts or desert margins.

The second image is of desertification. The term desertification conjures up images of advancing dunes, of ancient cities, once splendid, now half-sunk in seas of sand. This does happen, and when the deserts do move they do so with immense power.

Nevertheless, the shifting walls of sand are something of a rarity. Deserts do expand and contract somewhat, as satellite images graphically reveal, but usually oscillate within fairly well-defined ranges. Even the driest cities in the world are in no particular danger of being engulfed by sand.

More often, desertification of drylands on desert margins is something more insidious; and in contrast to the great sand migrations, it is often caused by people. This desertification is a triple tragedy: it impoverishes the people who live in the regions affected; it removes some small but significant areas of vegetation, important in the absorption of the greenhouse gas carbon dioxide; and it may narrow biodiversity.

The United Nations Environment Programme (UNEP) is committed to protecting the world's drylands from desertification. For 20 years, UNEP has undertaken an array of projects, from satellite mapping to working with desert nomads on management strategies. Too often, however, these programmes, and the programmes of other concerned organizations, have been slowed by a lack of funding. Deserts have not attracted the intense public attention that the rain forests have, and desertification has not caught the public imagination in the same way as have other ecological crises, such as ozone depletion or climate change. And all too often, governments fail to integrate fully the challenge of desertification into the development process.

This extraordinary publication should be read, referred to, or even just browsed through by everyone concerned with the fate of our planet. For in doing so, you show a concern for those people – largely voiceless – who call the drylands home, and whose home is under threat.

Mostafa K. Tolba
Executive Director
United Nations Environment Programme

FOREWORDS

The record of the rocks shows that there have been deserts on Earth for hundreds of millions of years. Throughout this time they have been constantly on the move in response to changing climates and drifting continents. Twenty thousand years ago there were forests and grasslands in the central mountains of the Sahara, and 8,000-year-old cave paintings show elephants, rhinoceroses and antelopes. Today, humans bear a burden of responsibility for much of the expansion of the deserts, as such things as overgrazing take their toll.

Desertification is not, however, a simple case of deserts expanding. For example, it is not true – as is sometimes thought – that the Sahara Desert has marched southwards into more fertile regions of Africa. In some semi-arid desert margins, agricultural productivity has increased in the past 20 years. Over the semi-arid regions as a whole, however, misuse of land has reduced fertility, and it has been estimated that 70 per cent of the world's dry rangelands are less productive than they were or should be.

The problems of desertification do not result from ignorance either of cause or remedy. They are due to the inability of human communities to apply what is known. Much of the knowledge is traditional. People over the centuries have evolved ways of living in desert areas in what we now call a "sustainable" way. But as populations have increased and the pressures to produce crops for export have

grown, traditional wisdom has been cast aside. And we continue to create new problems. In many dry regions, new irrigation techniques that deplete ancient groundwater are creating immense areas of unsustainable cultivation.

What can be done? This book allows us to see some of the answers. First, we have to understand the systems we are using and abusing. Second, we have to recognize that while bad management is all too often undermining the productivity of deserts, good management can enhance it.

Reversing the desertification caused by human activity will cost money, and this is in short supply. Only a pitiful 1 per cent or so of the funding called for to implement the 1977 Nairobi Plan of Action to Combat Desertification was ever received. In Rio de Janeiro in June 1992, many governments called for an international convention on desertification. This will also be useless unless the knowledge that the nations will share is backed by funds to apply it.

We have much to learn from the deserts as we look to the future of human civilization. The chief lesson we should heed is that nature is not infinitely tolerant, and that the amount of desert that tomorrow's generation will inherit lies very much in the hands of the people of today.

Martin W. Holdgate
Director General of IUCN

Deserts are one of the most extensive environments in the terrestrial world. The hot and cold deserts together cover almost 40 per cent of the Earth's surface; after the oceans, they are among the most important elements in the global climate system. (Although Antarctica is often regarded as being a desert, it is different in many important respects from the deserts we shall be discussing here and is not included in this book.) As well as being an influential part of the world environment, desert landforms in themselves are extremely active; massive quantities of boulders, sand, silt and clay are eroded, transported and accumulated by running water and the wind in desert regions.

With little or no vegetation, the world's deserts provide particularly striking vistas, whether viewed from the depths of silent space or from a vantage point on Earth, deep in a sand sea or on top of a barren mountain. The beauty of the deserts is only one aspect of their significance, but it has inspired the cultures of those who have lived there and captured the imaginations of those who have visited.

The rich diversity and sheer numbers of flora and fauna that are supported in many other ecological zones are not present in deserts. Nonetheless, the following pages reveal that the wide variety of adjustments made by the plants, animals and human societies to life in the deserts are extraordinary and in themselves highly diverse.

Since Neolithic times, and at an accelerating rate over the past 6,000 years, developing economic systems and ever more powerful technologies have enabled the peoples of the deserts to use and to distort natural processes. Technology played an integral part in the establishment of the first civilizations in very arid tracts of northeastern Africa, the Middle East and South Asia, mainly by harnessing the great rivers. More recently, technology has enabled access to formerly remote supplies of surface water and to deep groundwater. This water has augmented the scarce water resources of marginal tracts, thereby insuring agriculture against normal but damaging variations in rainfall. Groundwater has also enabled such things as the intensification of livestock rearing. The potentially degrading pressures that deserts and their margins are enduring are a major theme in this book.

The visual and verbal record of the world's deserts to be found in the pages ahead is exciting, but also disturbing. It demonstrates very clearly that many current practices are unsustainable. The need for economic and political adjustments in accord with sustainable principles is urgent, and the introduction of policies validated and implemented at national and international levels is essential.

Tony Allan
Andrew Warren

THE DESERT

Deserts are often imagined simply as large areas covered in sand; yet such regions make up only a small part of the deserts of the world. What, then, is a desert? Scientists have various definitions, but the essential features are a lack of water and a scarceness or lack of vegetation and other forms of life. This leads to many questions: what climatic conditions cause deserts, why some regions of the world are prone to desertification and others not, and whether some types of land are more likely to become desert than others. Deserts are not static; they have expanded and contracted, appeared and disappeared with changes in climate and landform. In the future, human activity and global warming may play a major role in determining the future of deserts.

Above *Dead trees in the Western Australian desert.*
Left *A solitary plant disturbs wind-ripples on a California sand dune.*

WHAT ARE DESERTS?

Deserts are the world's dry lands, places where there is little water, and often the water there is comes when or where it is of little use. Deserts are barren landscapes, where people, animals and plants can only rarely subsist. Life forms must develop a range of strategies to deal with water shortage.

Although it might be possible to point to a number of characteristics, the key issue in deserts is a lack of water, and scientific definitions have always considered the amount of water rather than temperature. The most widely used definition, developed by Peveril Meigs for UNESCO in 1953, balances water delivered by rainfall against water lost through evaporation and as vapour from the leaves of plants. This system classifies deserts as semi-arid, arid and hyper-arid. In all, these categories cover one-third of Earth's land surface. According to Meigs, arid zones are broadly equivalent to areas with average total annual rainfall of less than 200 millimetres (8 inches). Hyper-arid areas receive less than 25 millimetres (1 inch) annually, and semi-arid areas receive less than 600 millimetres (24 inches). There are virtually no places where rainfall is entirely unknown.

Not only are rainfall totals extremely low in deserts, but the rain that does come is unpredictable. Most deserts have no wet season, so that people, animals and plants cannot prepare with certainty for the arrival of the rains. Some areas may endure several years without any rain at all before one or two large storms suddenly deliver the entire rainfall for those years. Often the storms may be very local – a phenomenon that is sometimes called rainfall "spottiness". In the semi-arid desert margins, rainfall tends to be more seasonal.

Although considered inhospitable, deserts are home to 13 per cent of the world's population.

Many deserts, especially those located in the interiors of the continents, combine lack of moisture with high summer temperatures. Often, high temperatures are accompanied by great daily ranges of temperature, for deserts lack the cloud cover that acts to retain warmth over much of the rest of the Earth's surface. Deserts commonly experience daily temperature variations of 20°C (36°F), and some mid-latitude continental deserts – even the northern Sahara – have frequent night-time frosts in winter.

The climatic extremes found in deserts mean that plants have difficulty in establishing themselves. Vegetation cover is thin, with the result that only poor soils can develop. Many plants are adapted to take opportunistic advantage of sporadic rainfall by germinating from seed, growing and flowering very quickly once a storm comes. Others, such as the many species of cactus, are slow growing but efficient at moisture retention. Desert vegetation often grows in patches, or nodes, related to sources of moisture or nutrients, such as the sandy beds of ephemeral streams.

Desert landscapes

There is a great diversity of landscapes covered by the term "desert". Several scientific writers have divided deserts into two main structural types. The first consists of the great gently undulating areas formed by the break-up of the ancient "super-continent", Gondwanaland, between 100 and 200 million years ago. This first group includes most of the deserts of Africa, Arabia, India and Australia. The landscape is vast, barren plains, sometimes covered with sand, which can accumulate into fields of dunes.

The second type of desert is the mountain-and-basin desert. In these deserts, geologically recent movements of the Earth's crust have created mountain ranges, interspersed with alluvial plains and flats. The deserts of North and South America and of Central Asia fall into this category. On a more detailed scale, these two types may be further divided into river-formed landscapes, sand seas, clay plains, stony deserts, lake basins and upland areas.

Although deserts are considered inhospitable, around 13 per cent of the world's population live in the area covered by Meigs's classification. The hyper-arid cores are virtually uninhabited, and the greatest density of population lies in the semi-arid desert margins.

Top *A semi-arid landscape near Uluru (Ayers Rock) in the Northern Territory, Australia. Note the clouds which sometimes, but only very rarely, bring rain. This scant rainfall is sufficient to support a sparse vegetation of scrub and small trees, which consolidate the land surface into a series of "mats".*

Left *A massive sand dune advancing slowly (from left to right in the picture) across the Rub' al Khali (Empty Quarter) of southern Saudi Arabia. Although the sand will eventually smother all the vegetation on the plain beneath, the leading edge of the dune provides a temporary haven of moist and therefore favourable conditions.*

Above *Salt flats in Death Valley, California, USA, where the annual rainfall averages less than 50 mm (2 in). Salt flats are possibly the most inhospitable type of desert landscape, and are formed by the evaporation of lakes. The ridges that divide the surface of the flats into an irregular mosaic are pressure ridges formed at the margins of adjoining "plates" of crystallization.*

WHY ARE THERE DESERTS?

Deserts lie in areas where rainfall is low, or the loss of water to the atmosphere from evaporating water surfaces and transpiring vegetation is high, or there is a combination of these two. These dry areas are the result of one or more of four factors: atmospheric high pressure zones, "continentality", cold ocean currents and "rain-shadows".

High pressure zones

Most desert lie north and south of the equatorial rain forests around the 30°S and the 30°N latitudes, under the subtropical atmospheric high pressure zones within the "Hadley Cell" (a convection pattern driven by solar energy). Thus, African deserts are found in two bands – in the north the Sahara, and in the south the Kalahari and Namib.

High-pressure belts do not form continuous bands around the globe along the 30th parallels. Because of the distribution and the pattern of wind flow around landmasses in these regions, deserts tend to be driest on the western sides of the continents – prevailing winds tend to blow roughly easterly. Conversely, moist air travels around the high-pressure cells into areas which we might otherwise expect to be dry on the eastern sides of continents, such as the Gulf of Mexico in the western Atlantic and the Philippines.

The Hadley cell

The subtropical high pressure zones responsible for desert aridity are the result of a simple convection pattern called the Hadley Cell, which is driven by the Sun's energy. Because the Sun's radiation is greatest at the Equator, air is heated and rises, creating low pressure zones at the Earth's surface (1). Moisture-bearing air masses are then sucked into these zones, where rainfall is intense and rain forests occur (2). The rising equatorial air masses move towards the poles, and then subside as they cool (3). As the air nears the surface, it becomes drier and creates high pressure. These high-pressure cells, or "anticyclones", are known as subtropical highs. The air within is dry, and high pressure blocks the incursion of moist air. The air circulates back to the Equator (4).

Continentality

In the continental interiors, the sheer distance from the ocean prevents the penetration of moisture-bearing winds. This is particularly important in the mid-latitude deserts, located north and south of the belt of subtropical high pressure. The great deserts of Central Asia provide the best example of this kind of desert. The Takla Makan Desert of western China is the only hyper-arid desert outside the subtropical high pressure zones, although others outside these zones are, nonetheless, very dry.

Cold ocean currents

The third cause of deserts intensifies aridity rather than acting as its primary cause. Cold oceanic currents moving from the poles towards the tropics along the western coasts of the continents cool the sea and lower the rate of evaporation from its surface. This further reduces the already sparse available moisture. Notable examples are the Benguela Current along the southwest coast of Africa, which affects the Namib Desert, and the Peru Current along the west coast of South America, further drying the Atacama Desert. Seemingly paradoxically, these hyper-arid coastal strips are frequently engulfed in fog, which is created when cold sea meets warm air. In these areas, moisture is delivered more by fog than by rain. The fogs and sea breezes greatly moderate air temperatures, so that these hyper-arid deserts are actually areas of quite moderate heat.

Rain shadows

Dryness is intensified in some deserts that are sheltered from the prevailing wind by a mountain range. This is known as "rain-shadow": as air masses move up and over the mountains, moisture develops into clouds and falls as rain. On the other side of the range, the subsiding air is extremely dry. The best examples of this kind of desert are the intermontane basins of North America and Central Asia.

Classifying deserts

Deserts can be divided according to their rainfall conditions: some seldom receive rain or else experience fog instead of rain, others have either a winter or a summer rainy season, or even two rainy seasons. Temperature is also a useful means of classification. Some deserts are hot during the day all year, some have a great annual range of temperature, and others are cool and have cold winters. Scientists estimate that 43 per cent of arid lands are hot deserts, with average summer temperatures above 30°C (86°F), while 24 per cent are cold deserts, with average winter temperatures below 0°C (32°F).

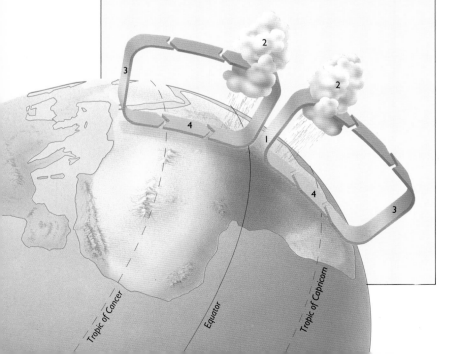

Right A map of the world's deserts clearly shows how most arid and semi-arid regions lie within the belts of high air pressure associated with the Hadley cell, in which the air is dry. Some deserts of Asia and North America lie in the interior of these continents, cut off from rain-bearing winds by mountain ranges.

Top right Sand dunes cover the Lekhwair oilfield in Oman. These dunes are an extension of the Rub al Khali (Empty Quarter), the great sand sea of southern Saudi Arabia. Although water is almost entirely absent, oil has drawn people into the desert. The Lekhwair field has been producing oil since the 1970s.

Tropic of Cancer

Equator

Tropic of Capricorn

Tropic of Cancer

Equator

Tropic of Capricorn

Warm ocean currents   Cold ocean currents   Winds

CLIMATE

Deserts are not all barren, windswept, sandy expanses where rain never falls. In fact, although all deserts are short of water, they exhibit a great range of climates. Scientific explanations of aridity concern a variety of atmospheric processes based on such things as sub-tropical high-pressure cells, continental winds, land forms and ocean currents. At any particular location, these combine to cause a particular climate regime.

Types of climate

Deserts have an immense range of air temperatures. In the Sahara, Arabian Peninsula, Sonoran Desert of California, Australian Desert and Kalahari, the anticyclonic (high-pressure) weather systems bring clear skies with high ground temperatures and a marked coolness at night. In tropical areas, like Somalia, there is little change in monthly temperatures. By contrast, deserts in continental interiors – notably the arid areas of Asia and the western United States – have large seasonal temperature ranges, with very cold winters and very hot summers. The Iranian, Nevada and Gobi deserts are sometimes described as having "temperate" climates because of their cold winters.

On the leeward sides of mountain ranges, such as the Sierra Nevada, the Great Dividing Range in Australia and the Andes, extremely arid conditions prevail. Any rain that falls from moisture-bearing winds does so on the mountains, and so there is a "rain shadow" effect.

Desert that have clear skies, as most do, receive large amounts of energy from the Sun. The high reflectivity of sand and rocks reduces the effects of this energy on the ground, but is not sufficient to prevent significant evaporation of water and heating of the air. At night, the clear skies allow rapid heat losses from both ground and air.

Wind speeds are not necessarily higher in arid lands. Indeed, because stable air masses are dominant, winds are often light. However, localized surface heating can produce high local wind velocities, as in dust devils.

Climatic fluctuations

The Earth's climate is neither constant nor stable, and has fluctuated significantly over both short and long periods. Desert conditions and the extents of deserts have fluctuated as a result. Of particular interest to scientists are the changes in deserts that took place with major global climatic changes, particularly during the ice ages, as well as the shorter-term fluctuations that have such profound impacts on the inhabitants of deserts.

The last one million years have seen ten major cycles of glaciation, or ice ages; the height of the most recent one was about 18–20,000 years ago. Drastic changes occurred in the Earth's water systems during these glacial periods. Much more water was stored in ice bodies and much less in the oceans. The effects of these and other changes extended beyond the areas directly affected, and had profound effects on the extent and location of deserts. Compared to today, temperatures during the last glacial were on average 9°F (5°C) less in the atmosphere and 2–3°C (4–5°F) less for ocean surfaces. These temperature differences meant that there was

less evaporation from the ocean surface which, together with the sheer increase in the area of land caused by the lower sea levels, contributed both to a reduction in rainfall and the expansion of arid regions in many parts of the globe.

As well as major expansions of desert conditions in association with climatic fluctuations, there have been occasions when wetter conditions penetrated into the heart of today's deserts, particularly at the height of the Holocene climate, 4,000–7,000 years ago. Archaeological remains – including rock paintings – and ancient pollen show that savannah vegetation occurred in parts of the central Sahara, supporting large human and wildlife populations.

Short term climatic conditions in deserts, particularly rainfall, are also subject to considerable fluctuations. It is not unusual for rainfall to vary by more than 50 per cent from year to year. Periods of drought, which can persist for several years, are a normal feature of desert conditions, as are the much shorter periods when above-average rainfall occurs. In parts of Sudan, for example, annual rainfall from 1965 to 1985 was 40 per cent less than from 1920 to 1940.

Above *Travellers in Niger, West Africa, seek the meagre shelter of thorn bushes in the face of a rainstorm. The most typical features of desert rainfall are its sudden ferocity and its unpredictability. Although in some places there may be distinct "wet" and "dry" seasons, a storm can break at almost any time, with little in the way of advance warning.*

Inset above *An aerial photograph showing early morning fog over sand dunes in the Namib Desert of southern Africa. Such fogs are the Namib's main source of water. Moisture-laden air blown in from the sea cools and condenses when it comes into contact with the night-chilled desert surface. Many plant and animal species use the fog as a supply of water.*

Left *Death Valley, California, is one of the hottest places in the world. Desert skies generally have some clouds, but it is rare for them to gather in sufficient density for rain to fall. The dry water courses in the foreground, however, show that the effects of precipitation are not completely absent.*

DESERTS OF THE PAST

The world's deserts are in a constant state of change. Because climatic conditions are the reason for the existence of deserts, and these conditions are subject to change over a wide range of time scales, the extent and character of deserts have changed too. For the Earth as a whole, the unequal distribution of heat from the Sun – with most in the tropics and least at the poles – has driven the atmospheric circulation and ocean currents in much the same way for much of the planet's history, with the result that arid areas have existed for a similarly long period. These ancient deserts have left their mark in the geological rock record.

The record is incomplete, for several reasons. One of the reasons is that deserts embrace a range of environmental conditions, and depositional environments – where the sediments that form rocks accumulate – are just one. So the rock record is only likely to preserve part of the story. Furthermore, sediments formed under desert conditions in the past have themselves been subjected to erosion, deformation and reworking.

In the past, most deserts were generally in the subtropics in both hemispheres because of the prevailing atmospheric conditions, as is the case today. The movement of the continents, however, now means that ancient desert rocks can today be found elsewhere on Earth.

Deserts in the rock record

If it is possible to identify the processes responsible for the formation of specific sediment characteristics today, and to find the same characteristics in ancient rocks, it can be assumed that the ancient rocks were formed by similar processes under similar environmental conditions.

One of the first descriptions of ancient deserts in the United States came from sandstone exposed in the walls of canyons in Utah. In 1907, Ellsworth Huntington used the structure of the stratification to attribute the origin of this sandstone – known as the Navajo Sandstone – to deposition as sand dunes under desert conditions in the Jurassic and Triassic periods (150–200 million years ago). Since then, many such sandstones have been found. The structure of the rocks has allowed the actual types of dune to be determined. In Great Britain, for example, the New Red Sandstones of the Midlands are known, by their structures, colour and fossil content, to be derived from desert dune deposits, formed when Great Britain was located nearer the Equator. Ancient desert wadi flood deposits and evaporites, containing high salt concentrations, have also been identified, sometimes in close association with ancient dune sands, as in the case of the Rotliegend desert rocks of the North Sea basin, which date from the early Permian period (about 290 million years ago).

Today, identifying ancient desert sandstones is economically important. Their porosity has made them major reservoirs for oil and gas. The ancient dune sands of Colorado and Wyoming, the Cooper Basin of South Australia, and northwest Europe are important examples of ancient desert sediments that are now major oil sources.

Deserts in the Quaternary period

Some landforms and sediments on the surface, rather than in the rock record, provide information about the extent of deserts during the more recent past, particularly the Quaternary period (the last 2 million years). Perhaps the most widespread indicators of Quaternary deserts are the degraded desert sand dunes that occur in many parts of the world, including North and South America, Australia and India. In Africa, areas bordering today's Sahara and Kalahari possess sand dunes that are far beyond the limits of present-day desert conditions. In northern Matabeleland in Zimbabwe, scientists have identified the remnants of extensive former dune systems, now covered with thick forest. In Nigeria, old dunes have been gullied and soils have developed on their surfaces. In many cases it is still not possible to determine accurately the age of these relic features, though many have been attributed to the last ice age (18–20,000 years ago), when dry conditions were probably more widespread.

Arid areas have existed throughout the planet's history.

Above *The barren landscape of the Colorado Plateau in the western United States consists of the Navajo Sandstone eroded into pavements and conical outcrops. This sandstone was deposited as desert sand dunes between 200–150 million years ago.*

Inset *Fossil sand dollar (sea urchin) found in the Kavir Desert, Iran. Such fossils prove that the now-arid Kavir was once the bed of an ancient sea.*

Right *Erosion caused by wind and rain has exposed multicoloured rock strata in the Painted Desert in Arizona, USA. These alternating layers of sand, clay and shale were laid down more than 70 million years ago. Elsewhere in the Painted Desert is found the Petrified Forest which contains the fossil remains of giant prehistoric trees that grew about 225 million years ago.*

WATER IN THE DESERT

Rainfall is not completely absent in desert areas, but it is highly variable. An annual rainfall of 100 millimetres (4 inches) is often used to define the limits of a desert. The impact of rainfall upon the surface and groundwater resources of the desert is greatly influenced by landforms. Flats and depressions where water can collect are common features, but they make up only a small part of the landscape.

Arid lands, surprisingly, contain some of the world's largest river systems, such as the Murray-Darling in Australia, the Rio Grande in North America, the Indus in Asia and the Nile in Africa. These rivers and river systems are known as "exogenous", because their sources lie outside the arid zone. They are vital for sustaining life in some of the driest parts of the world. For centuries, the annual floods of the Nile, Tigris and Euphrates, for example, have brought fertile silts and water to the inhabitants of their lower valleys. Today, river discharges are increasingly controlled by human intervention, creating a need for international river basin agreements. The filling of the Ataturk and other dams in Turkey has drastically reduced flows in the Euphrates, with potentially serious consequences for Syria and Iraq.

The flow of exogenous rivers varies with the season. The desert sections of long rivers respond several months after rain has fallen outside the desert, so that peak flows may be in the dry season. This is useful for irrigation, but the high temperatures, low humidities and different day lengths of the dry season, compared to the normal growing season, can present difficulties with some crops.

Regularly flowing rivers and streams that originate within arid lands are known as "endogenous". These are generally fed by groundwater springs, and many issue from limestone massifs, such as the Atlas Mountains in Morocco. Basaltic rocks also support springs, notably at the Jebel Al Arab on the Jordan-Syria border. Endogenous rivers often do not reach the sea, but drain into inland basins, where the water evaporates or is lost in the ground.

Most desert stream beds are normally dry, but they occasionally receive large flows of water and sediment. These flash floods provide important water supplies, but can also be highly destructive. Floods are discussed on pages 40 and 41.

Groundwater resources

Deserts contain large amounts of groundwater, when compared to the amounts they hold in surface stores such as lakes and rivers. But only a small fraction of groundwater enters the hydrological cycle – feeding the flows of streams, maintaining lake levels and being recharged through surface flows and percolating rainwater. In recent years, groundwater has become an increasingly important source of fresh water for desert dwellers. The United Nations Environmental Programme and the World Bank have funded attempts to survey the groundwater resources of arid lands and to develop appropriate extraction techniques. Such programmes are much needed, because in many arid lands there is only a vague idea of the extent of groundwater resources. It is known, however, that the distribution of groundwater is uneven, and that much of it lies at great depths.

Groundwater is stored in the pore spaces and joints of rocks and unconsolidated sediments, or in openings widened through fractures and weathering. The water-saturated rock or sediment is known as an "aquifer". Because they are porous, sedimentary rocks, such as sandstones and conglomerates, are important potential sources of groundwater. Large quantities of water may also be stored in limestones when joints and cracks have been enlarged to form cavities. Most limestone and sandstone aquifers are deep and extensive, but may contain groundwaters that are not being recharged. Most shallow aquifers in sand and gravel deposits produce lower yields, but they can be rapidly recharged.

Some deep aquifers are known as "fossil" waters. The term "fossil" describes water that has been present for several thousand years. These aquifers became saturated more than 10,000 years ago and are no longer being recharged.

Water does not remain immobile in an aquifer, but can seep out at springs or leak into other aquifers. The rate of movement may be very slow: in the Indus plain, the movement of saline groundwaters has still not reached equilibrium after 70 years of it being tapped. The mineral content of groundwater normally increases with the depth, but even quite shallow aquifers can be highly saline.

Left An arid landscape near Lake Mead in Nevada, USA. The dry river bed (running to the left foreground) carries water only during rain storms. The water nonetheless has enough power to clear debris. Without a concealing layer of vegetation, every detail of the river bed and its tributaries is revealed in bold and sharp relief.

Above The Niger River at Mopti in Mali, West Africa. The Niger rises among rain-forested mountains about 200 km (120 miles) from the coast, then loops more than 2,000 km (1,200 miles) inland, flowing through the southern margins of the Sahara Desert. The Niger is a permanent resource for water, food (in the form of fish) and transport.

DROUGHT

Above *Refugees from drought at a relief camp in Sudan, East Africa.*

The idea of drought – a long period without rain – is simple to understand, but hard to define in a useful way. If a drought were simply a shortage of rain, the Sahara would be in permanent drought and the Sahel would be considered well watered. It is better to regard a drought as being "less rain than expected", and to focus on how people come to be disappointed in their expectations of rain.

Droughts happen erratically, which makes planning very difficult. For example, a dry spell lasting over 20 years in the Sahel has had two "dips" – in the early 1970s and in 1984 – when the rainfall was even lower than for the rest of the time. It is difficult to decide whether the long period was simply a drought or evidence of a more long-term climatic change.

Drought may occur in many parts of the world. There were severe droughts in Australia in 1902, 1912–15, 1965–67 and 1972; the worst drought for two centuries began in 1983. In the early 1990s, northeast Brazil had its worst drought since 1583, and southern Africa had a terrible drought in 1992.

There will be more droughts, long and short. Although meteorologists do not know when they will arrive, some clues are emerging. The Biblical theory of seven lean years and seven fat years suggests that there is a regular drought cycle. This may prove to be partly correct. However, cycles of drought, if they are shown to occur, will be much more complicated than a regular seven-year pattern.

The distress that droughts bring is often the result of misplaced expectations. By the end of the 1960s, a decade of better-than-usual rains in the Sahel had persuaded people to plant crops and settle in areas that had previously been too arid. Populations grew rapidly and people saw a modest increase in their standards of living. Their expectations rose, and this led in turn to a greater demand for resources. When the rains failed, the drought had a far greater impact.

The idea of drought gets complicated if we look in more detail at the problems of survival in the desert margins. In wet years, for instance, hollows may be waterlogged and produce poor crops; in the dry years they may collect the only water there is, and produce the best crops. The soil type of an area affects how quickly water evaporates in dry spells. The impact of a drought also depends on the types of crop that are grown.

Coping with drought

People in the dry lands have to learn to live with drought by careful management of resources. Drought is now a matter for national and even worldwide concern. Since the terrible African famines of the 1970s and 1980s, when thousands died as a result of drought, there have been concerted efforts to prepare for the inevitable scourge. Scientists now use satellite images to measure the progress of rains, and so to identify droughts before people begin to starve. This method gives accurate warnings, but very often aid agencies lack the resources to respond in time. Other strategies to alleviate drought include improvements in transport and storage so that food can quickly reach the needy. Despite all this, catastrophic droughts hit parts of the Horn of Africa in the 1980s. The effects of these were magnified by population growth and the dislocation caused by war.

THE ARID LANDSCAPE

The most striking thing about desert landscapes is their stark beauty, so different from the rich terrains inhabited by much of the world's population. Some deserts appear to be little more than flat scrubland, yet others are filled with extraordinary forms sculpted by wind, sand and water. The single most important factor in a desert environment is water, or rather the lack of it. The dryness allows sand and dust to be whipped up into storms that rival the ferocity of any blizzard. When rain does come, floods may sweep across the desert floor. Yet, despite their apparent ferocity, desert environments are ecologically delicate and can easily be thrown out of balance by various human activities.

Above *The "Walls of China" around Australia's Lake Mungo.*
Right *Scoured by sun and water, a sterile landscape in the Negev Desert.*

DESERT LANDFORMS

From the deep canyons of the southwestern United States and the huge alluvial fans buttressing the Andes to the vast salt flats of the Sahara and the huge areas covered by regular ridges of sand in Arabia, the diversity of desert landforms in both scale and origin is breathtaking. Arid environments range from dry lowland basins – where strong winds mould dunes and carve rock – to erosional badlands and barren precipitous mountains, where the action of water dominates the creation of landforms. The vast array of different landscapes and landforms in arid lands is a result of complex differing balances of the various processes that are at work in desert environments.

The blanketing of a quarter of the world's desert surfaces in wind-blown sand suggests that winds are very significant in moulding the desert surface. Hundreds of square kilometres of dunes, most over 3 metres (10 feet) in height, each progressing at 15 metres (50 feet) every year, is not an unusual occurrence. In central China, wind-eroded hills, or *yardangs*, covering about 10,000 square kilometres (4,000 square miles) are testament to the power of the wind.

The term sand is applied to rock fragments with a particle size between 0.2 and 2 millimetres (0.008–0.08 inch). Desert sand grains are on average 25 times larger than particles of sedimentary clay. Wind-borne sand grains can be identified under the microscope by the frosting of their facets caused by multiple impacts; they are also usually well graded by size when they are deposited.

Another feature created by the wind is the desert pavement, known as *reg* in the Sahara. *Reg* is formed by progressive wind erosion of a bedrock matrix, leaving behind pebbles of more resistant rock, such as chert and flint. These pebbles become very closely packed on the surface and greatly reduce subsequent erosion of underlying material.

Wind is more effective in deserts than in other regions, not particularly because it is any stronger than in wetter environments, but because dry surfaces and a lack of protective vegetation make it more efficient. Despite the dominance of wind in the creation of landforms in the very arid areas of the globe, it is surpassed, as an agent of erosion, on the desert margins and in the highland regions by water.

Above *Salt pillars rise above the highly saline waters of the Dead Sea in Israel, the lowest point on terrestrial Earth. The only perennial inflow, from the River Jordan, is fresh to taste, but intense evaporation over millennia has concentrated the salt in the water. Without inflow, the Dead Sea would evaporate away completely within a relatively short space of time.*

Left *Eroded rock formations dominate the landscape of the Pinnacles Desert In Western Australia. The combined forces of wind and weathering produce these low, conical outcrops that are formed out of slightly more resistant rock. In the foreground, the rock is visibly breaking down into the sand that covers the desert surface between the rock pinnacles.*

Far left *The Totem Pole rock formation overlooks wind-rippled dunes in Monument Valley, Arizona, USA. Although the Totem Pole is an extreme example, its shape is typical of desert rockforms in this region. The slow retreat of cliffs has left behind pillars of what were once much broader tablelands.*

Rainfall and river-flow in desert regions are not continuous through the year, but when streams and rivers do fill with water, they have tremendous power to erode and carry sediment. This power is exhibited by deep erosional gorges, such as the Grand Canyon, as well as by the huge alluvial fans which are deposited as narrow channels in the mountains and open out into the lowland plains.

Few of the landforms of the desert are produced solely by the action of either wind or water. Commonly, the two major agents of erosion and deposition work in collaboration. Most of the sediment worked by the wind into dunes was first eroded and accumulated by water. Similarly, much of the water-borne material is derived from erosion of previously wind-blown sediment.

Other processes are also at work in desert environments which are not as conspicuous as in less arid areas. One of the more important is weathering. The daily extremes of heat and cold in many desert areas can crack and degrade rock surfaces. In some high-altitude deserts, such as the Karakoram of Pakistan, frost shattering of rock can occur. Similarly, rock disintegration can occur from the growth of salt crystals in rock joints. These weathering processes produce vast amounts of sediment, which are then worked by wind and water. They can also produce bizarre landforms, like towers of resistant rock, or inselbergs, emerging from a flat plain, such as Uluru (Ayers Rock) in Australia.

Perhaps the most unexpected of desert landforms are salt flats or salt pans. These result from high rates of evaporation and are neither the product of "saltation" (the low-level effect of wind-blown sand) nor "salination" (an increase of salt in soil caused by irrigation). Salt pans are the residue of large bodies of water that have evaporated. Some, such as the Great Salt Lake in the United States, still have some surface water. Others, such as the Kavir of central Iran, are completely covered by a salt crust. The salt prevents further evaporation of water, and as a result the surface crust may conceal boggy ground beneath. These salt marshes are known as *sabakha* in the Arabian Peninsula.

SCULPTED BY WATER

Despite the limited rainfall, running water has played a significant role in shaping desert landscapes, eroding highland regions and depositing sediment in the lowlands. Sparse vegetation and thin soils, coupled with steep slopes and rainstorms of short duration but high intensity, mean that when water does come, it is a powerful agent of erosion and the transport of material. Indeed, some semi-desert environments produce the highest rates of erosion to be found anywhere on Earth.

Erosion of the land

Some of the most distinctive desert environments are the badlands – barren "moonscapes", devoid of vegetation and dissected by dense networks of rills and deep steep-sided gullies. These landscapes are common in western North America – particularly on the Great Plains and Colorado Plateau – as well as in Mexico, China and Spain, although they cover less than five per cent of desert areas. They form naturally on soft shale rock, and where the land is stripped of plant cover by grazing animals or poor farming.

Erosion rates in badlands are extraordinarily high, varying from 2 to 20 millimetres (0.08 to 0.8 inches) each year, despite the fact that rainwater runoff is only periodic. The rate of erosion may be so rapid that formation of new soil is impossible. This prevents vegetation becoming established, which would help to stabilize soil, and erosion becomes more and more severe. The arid climate further hinders the growth of vegetation. Erosion occurs both on the surface as "sheet-wash" and beneath it as "piping" or "tunnel" erosion. Tunnel erosion occurs when rainwater seeps into deep cracks in the soil. The cracks are enlarged and elongated into tunnels, some of which may be quite large – as much as 30 metres (100 feet) long and 2 metres (6 feet) wide or more.

The majority of badlands seen today are probably not contemporary features. The formation of those in New Mexico and in Israel's Negev Desert were triggered by changes in the Earth's climate that occurred about 40,000 to 70,000 years ago. Many of the world's other badlands probably began forming in response to global climate changes thousands of years ago.

Deposition by water

Short-lived, fast-flowing streams acquire a heavy load of sediment as they rush down from highlands. When they leave the confines of hills and reach flat, wide desert plains, they spread out and lose their power to carry sediment. Large amounts of cobbles, sand, silt and clay are then deposited between mountain and plain. As a result, these environments are characterized by alluvial fans – cones of coarse sediment dumped on mountain slopes by overloaded streams.

Alluvial fans are characteristic of arid and semi-arid areas (although they also occur in other environments) because a high rate of sediment supply is required for their formation. They can be derived from single or multiple stream sources and can coalesce into large fan complexes. The streams that deposit the sediment frequently wander across the surface depositing on different parts of the fan. Commonly, because

of some past environmental change, the stream will cut down into its own deposits, leaving relic fan surfaces at various heights. Alluvial fans can take a long time to form and their deposits are therefore an important record of past environmental change. For instance, the Milner Creek fan in the White Mountains of California is still growing after 700,000 years. Alluvial fans are also important for groundwater resources, and have potential for agricultural development (the silts also contain nutrients) and as sites for settlement. However, they are subject to flash floods and unpredictable erosion and deposition, making human activity hazardous. Some flash floods on alluvial fans occur as debris flows. These are thick, slow-moving rivers of mud made up of a mass of debris which advances downslope in waves behind a front of rolling boulders. The effects of such floods can be very dramatic and very destructive.

The short-lived streams, or wadis, of the desert are also a part of the depositional process. A wadi typically has a wide, flat bed, which may be dry for much of the year. There may be scattered vegetation growing in the bed, sustained by groundwater or the infrequent flows. Despite their ephemeral nature, wadis can transport vast quantities of sediment for considerable distances across the desert.

Left *Dead Horse Point on the Colorado River in Utah, USA. Fed by mountain waters, the river has carved down into the desert surface through rock strata that represent hundreds of millions of years. The flat terraces indicate variations in the river's course through time. Secondary erosion by local groundwater has created the steep V-shaped gulleys in the sides of the main river canyon. Most of the resulting debris has been carried away downstream.*

Below *Ancient silty dunes, now dissected by water erosion, form the "Walls of China" near the dry Lake Mungo in southeastern Australia. Deposited during an earlier climatic phase, the dunes have now suffered a change in climate. Their saltiness prevents colonization by plants, leaving the surface bare and eroded.*

Below *Hickman Bridge, a natural stone arch in Capitol Reef National Park, Utah, USA. Both the arch and the low vertical cliff in the middle distance were carved by the action of a river that flowed sometime during the last few million years and which has since dried up.*

SEAS OF SAND

Left *Remarkably straight ripples formed by a wind blowing towards the camera, furrowing the early-morning surface of a dune. The dune itself was formed by a stronger daytime wind from the right. The stronger wind is just re-establishing its control and blowing sand off the tops of the ripples.*

Sand dunes cover about 20 per cent of the total area of the world's deserts and a much higher proportion of certain deserts such as those in Arabia. Most dunes are gathered together in vast areas termed "sand seas", or *ergs* in Arabic, where winds continually drift and mould the loose sand into dunes of various scales and forms.

The largest sand seas are found in North Africa, Arabia and the desert basins of Central Asia. There are also isolated examples in Australia and Namibia. The greatest number of sand seas occur in the Sahara, but the largest single sand sea is the Rub Al Khali (Empty Quarter) in Arabia, which covers an area of 560,000 square kilometres (216,000 square miles).

Sand seas accumulate over thousands of years in basins bordered by mountains. The basins act as traps for wind-blown sediment, which comes ultimately from weathered rock by way of alluvium in river valleys, exposed lake deposits or beaches. The deposits of the Great Eastern Sand Sea in Algeria probably took more than 10,000 years to accumulate and, if spread out evenly, would be 43 metres (140 feet) thick.

In most sand seas, only about 60 per cent of the area is actually covered with dunes. The duneless areas may be sand sheets (undulating plains of sand) or stony desert pavements. The dunes themselves exhibit a vast array of types, ranging from small ridges less than 1 metre (3 feet) high, to huge star-shaped dunes, such as those found in the deserts of China and Algeria, which can be more than 365 metres (1,200 feet) high and over 800 metres (2,600 feet) across.

Dune formation

The simplest dunes accumulate in the lee of bushes, large rocks or hills, such as the Draa Malichigdune in Mauritania, which is 100 kilometres (60 miles) long. Others are formed by the accumulation of sand in shallow surface depressions, or from sudden bursts of strong wind which sweep sediment into small piles. Once a mound of sand has formed, it will tend to replicate itself downwind as a result of the way that it disturbs the airflow. Thus, sand dunes often occur in assemblages of regular pattern and similar height.

Dunes of different types occur because of differences in annual wind pattern and sand supply. Where the wind blows in one direction throughout the year and sand is plentiful, sinuous ridges are formed at right angles to the wind. If sand is scarce, the ridges break down into isolated crescent- or horseshoe-shaped dunes known as "barchans". These are a relatively rare type of dune, but are well known because of their speed of movement. Sand eroded from the gentle windward slope of the barchan dune is deposited on the steep lee face and the dune "rolls" downwind. The speed of movement is governed by the size of the dune. Small barchans less than about 1 metre (3 feet) high may move 50 metres (150 feet) per year, although a more common figure for an average 10-metre (30-foot) high barchan is 5–10 metres (15–30 feet) per year. Barchan dunes do not grow in height indefinitely but tend to reach an equilibrium size which depends on environmental conditions, such as wind strength.

Linear dunes

The most common dunes are long, sinuous linear or "sayf" dunes, characteristic of the Namib and Australian dunefields. Their shape and direction are governed by a combination of winds coming from different directions in different seasons. Some of these dunes may reach as much as several kilometres (miles) in length and 170 metres (560 feet) in height. They commonly occur lying parallel to each other and give the landscape a striped, "corduroy" pattern. Unlike barchan dunes, these linear dunes are not very mobile and as a result they can offer a habitat in which vegetation can become established. Thick belts of shrubs and grasses frequently cover the stable regions at the base of linear dunes.

Unstable and stable

In highly unstable wind regimes where the wind direction is variable throughout the year, massive star dunes accumulate. These dunes are pyramidal in form, with long, sinuous arms radiating from the summit. They become sand traps, accumulating mass and losing little sand to the surrounding desert plains. They cover large parts of the Great Eastern Sand Sea in Algeria, where they occasionally reach 1,000 metres (3,300 feet) wide and 300 metres (990 feet) high.

Some dunes on the margins of deserts are not active in today's environment. They are relics of past, dry periods and are now stabilized by vegetation. Many were active at the height of the last ice age (18,000 years ago).

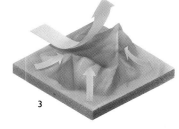

Above *Dunes in the Great Eastern Sand Sea in the northwestern Sahara. These dunes have been built up by winds from the southwest in winter and the northeast in summer. The sharp crests are small "reverse slip faces" and on the photograph they show that the wind has changed, probably during the day. The biggest dunes here are probably several thousand years old.*

Inset above *A massive dune towers above the Sossus Vlei in the Namib desert of southern Africa. The few thorn trees will do little to halt the slow advance of this huge dune. It is probably moving at less than 1 metre (3 feet) per year.*

Left *The relationship between wind (shown by arrows) and three types of dune. A barchan (1) formed by wind from one direction all year. Linear dunes (2), the most common type, are puzzling, and the diagram shows just one of various theories. Star dunes (3) are much less common and seem to form where winds are very variable.*

WIND EROSION

The power of the wind to erode, or alter the landscape, has long been underestimated. Wind erosion is effective not so much from the power of the wind itself, but from abrasion by sediment that it carries. In deserts, the high winds and the abundance of loose material makes wind erosion a very powerful force. This power manifests itself in a wide variety of wind-sculpted landforms, from streamlined hills to sandblasted stones, and also in the localized erosion of surfaces normally covered by vegetation.

Processes and products

Within about 40 centimetres (16 inches) of the desert surface, the wind transports sand grains through a process known as "saltation". The sand moves by bouncing along the surface at a height and speed governed by the strength of the wind and the size of the particles. The high-velocity impact of the sand grains in this near-surface region is an extremely efficient form of erosion. Stones and rocks on the desert surface are grooved and polished by this sandblasting process, and are shaped into ventifacts (wind-faceted stones). These can be identified by smooth faces, which intersect at sharp angles. The limited height of the "saltation curtain" means that this process of erosion is most efficient below 20–30 centimetres (8–12 inches). Evidence for this comes from the undercutting of rock at this height.

On a larger scale, the effect of wind erosion can be seen in wind-sculpted hillocks, generally known as yardangs. These are hills that have been abraded by the wind into streamlined, tapered forms. They occur where the wind comes generally from one direction throughout the year. Vast fields of yardangs aligned parallel to the wind can be found on the Peruvian coast, in Iran and in the Western Desert of Egypt. Yardangs range in size from as small as a metre (yard) in length, to up to several kilometres (miles). Some of the best examples are found around the Tibesti Plateau in the Sahara Desert, where parallel arrays of yardangs, spaced about one kilometre (three-quarters of a mile) apart, cover a total area of 650,000 square kilometres (260,000 square miles). These forms can reach 200 metres (660 feet) in height, and many kilometres (miles) in length. Most yardangs are carved from soft, loose rocks, but the Tibesti yardangs are cut in hard, ancient sandstones. Yardangs occur most commonly in corridors of sand movement where high winds hurl the sediment against the rock faces. On soft rocks, the rate of erosion can reach as much as 2 centimetres (0.75 inch) each year. The Tibesti yardangs are therefore likely to be hundreds of thousands of years old.

Powerful localized wind erosion can take place when a surface has been weakened by a break in the crust or by the removal of vegetation. In the Kalahari Desert and Great Plains of the United States, scouring by the wind – a process known as deflation – has created large semicircular, elongated hollows, up to a kilometre in diameter, which are devoid of vegetation. The deflation of surface materials by wind is a serious problem for agriculture in semi-arid areas. In the arid Sahara, where there is no vegetation to temper deflation, vast erosive plains have been formed.

Above Rock pillars carved by wind and weathering on the Tassilli Plateau in southern Algeria. Although rain and frost have also played a part, it is mostly the wind that has shaped this landscape, scouring the rocks and carrying away the loose material. Over many thousands of years, the erosion of the Tassilli has produced the great sand seas of the central Sahara.

Right Wind-sculpted rocks are a common feature in the world's sandy deserts. These pillars are part of the deeply wind-furrowed landscape of the central Sahara. They were carved by the abrasive action of sand-laden winds, some of which can be very strong. When bedrock becomes furrowed in this way on a large scale, the landforms are known as yardangs.

SOILS

Desert soils have four distinguishing characteristics. First, large quantities of dust are continually falling on them. This dust usually contains different constituents from the rock debris with which it mingles. Second, because they are less moist, desert soils are less chemically altered than humid soils. Third, not being leached continuously with water, they are generally saltier than humid soils (sometimes very much more so). And finally, because chemical alteration, leaching and erosion take place so slowly, desert soils have retained many more features from the past.

The uppermost layer on many desert rock surfaces is a thin, dark surface patina known as "desert varnish". Most desert varnish is very old, but some has formed fairly recently. Ancient petroglyphs (carvings or drawings on rock) in Australia and the southwestern United States have been found shrouded in desert varnish.

Early geomorphologists thought that desert varnish had appeared by "sweating": evaporation brought water to the surface, carrying iron and manganese from underlying rock layers. Later, scientists found that the amount of these elements in desert varnish was too great to have come from the rock layers alone, and that most in fact came from dust. It is not yet certain how the elements are fixed, but scientists have found evidence that in some desert varnishes, elements are scavenged and fixed by lichens and bacteria.

Outside of the great sand seas and zones of bare rock, the topmost layer of most desert soils consists of angular stones, known as the "desert pavement". This layer is seldom more than one stone thick, with far fewer stones lower down. The earliest explanation for this layer suggested that the finer soil between the stones had blown away. But while some experiments have shown that fine soil has been washed away by occasional showers, others suggest that stones edge their way to the surface through the expansion and contraction of the soil. The most recent explanation asserts that dust falling on the surface may then be washed beneath the stones.

Sub-surface layers

The fine soil beneath the desert pavement may rest on a number of sub-surface layers accumulated from soluble salts washed down into the subsoil. In humid soils, these salts are washed out entirely. In deserts, there is a sequence of these accumulations. Very dry desert soils, like those in the Egyptian Sahara, have layers of common salt (sodium chloride). In slightly wetter areas, such as southern Tunisia and parts of New Mexico, the common salt is washed out and gypsum (hydrated calcium sulphate) is washed down to form a soil "horizon". On the wetter edges of the desert, as in most of the Mediterranean, large swathes of Australia and in much of the southwestern United States, the characteristic horizon is formed by calcretes (calcium carbonate), common salt and gypsum having been leached away.

Some desert soils have other hard sub-surface horizons, which date from a wetter past. These horizons include laterites (rich in iron) and silcretes (rich in silica). In Australia, silcrete horizons contain gem-quality opal. Like calcretes, laterites and silcretes are as hard as rock.

River valley soils

In the lower parts of desert landscapes, there are two chief types of soil. In some hollows in semi-arid lands – where there is not too much salt – silica washed down in drainage water combines to form clays that hold organic matter tightly together. These black clay soils are known as "vertisols". They also form through the breakdown of basalt rock in semi-arid conditions. They are very fertile when irrigated and, for example, support cotton growing in the Deccan Plateau of India and in the central Sudan.

The other type of soil is far from fertile. In the silt deposited by rivers or streams, salt may be brought to the surface by capillary action from a water table a few metres below the surface. When the water evaporates, the salt remains, and can build up to form a white surface with conspicuous salt crystals. This kind of salinization occurs at alarming rates in many irrigation schemes, where the water table has been brought too close to the surface by heavy applications of water. In ancient Iraq, salinization caused by large-scale irrigation practices brought an end to many early civilizations. Today, salinization is active in Pakistan, to give just one example, and is a very real danger in any irrigation scheme in the dry parts of the world.

Above *Desert varnish on rock in the Sinai Desert. The coating of varnish takes many thousands of years to form. Desert varnish has been found on ancient rock carvings, for example in the Australian desert. The fact that the varnish loses some of its constituent substances more quickly than others helps to date these artefacts.*

Inset left *A desert pavement in the Algerian Sahara. The upright stones have been edged up as the soil expanded and contracted when it was heated and cooled or when it was wetted and dried.*

Left *A pebbly surface in the Namib Desert of southwestern Africa. Many desert soils have this kind of surface, created partly when the wind removes fine grains, partly when occasional storms wash grains away, and partly when pebbles rise to the surface as the soil is intensely heated and cooled.*

Above *A desert (or sand) rose from the Sahara. This flower-like form is an aggregate of gypsum and reddish coloured sand grains. Desert roses are thought to form below the surface, "growing" from the upper boundary of the gypsum horizon. Removal by the wind of the surface sand exposes the desert rose at the surface. The rosette shape is a result of the orientation and inclination of individual gypsum crystals. Desert roses are often found in Morocco and Tunisia, and Arizona and New Mexico in the United States.*

DUST

Huge amounts of dust are produced in deserts by weathering and abrasion by sand carried in the wind. Even more dust originates from sediments deposited by rivers, and the actions of people now play a large part in dust generation.

The biggest sources of dust are in arid and semi-arid environments where there are large areas of bare, loose sand and silt accumulated in alluvium (a fine silt deposited by rivers in floodplains and estuaries), dry lake beds and salt pans. The Sahara Desert produces somewhere between 60 to 200 million tonnes (tons) of dust per year and is believed to be the world's largest source of dust. The source area for a single duststorm is usually quite small, but when the dust is carried by high winds, a duststorm can expand from a few small plumes to cover a vast area. During the Shamal season in Iraq, dust plumes over the Arabian Gulf extending more than 400 kilometres (250 miles) have been recorded. These plumes may only be 10 kilometres (6 miles) wide at their source, but once carried downwind they can easily reach over 60 kilometres (37 miles) in width.

Dust tends to be generated when wind and water act together. Hence, hyper-arid areas tend not to be as dusty as semi-arid areas where there is some annual rainfall. The two dustiest places on Earth are both associated with the accumulation of water-borne sediment. They are the Seistan Basin in Iran and the floodplain of the Amu Darya in the south of the former Soviet Union near Afghanistan, which have over 100 duststorm days every year.

Long-distance dust

Once caught up in air currents, dust from deserts can be transported over many thousands of kilometres (miles). The deposition of red dust from the Sahara by rainfall in the British Isles ("red rain") is a fairly common occurrence and just one example of the phenomenon. There are numerous other long-distance dust routes, such as from West Africa to Miami, southern Ukraine to Sweden, and Mongolia to Hawaii. The summer dust haze that appears in the Arctic has been attributed to Central Asian desert sources, and dust from storms in the Great Plains and Canadian Prairies is often deposited more than 3,000 kilometres (1,800 miles) away.

Dust storms are highly seasonal in occurrence. They generally happen when the wind is strongest (for example, July–August in Arizona, March–May in Saudi Arabia). In cultivated semi-arid areas, such as many parts of the Great Plains of North America, duststorms occur most frequently in spring after the fields have been ploughed, when they are bare of vegetation and vulnerable to erosion.

The Sahara Desert produces somewhere between 60 to 200 million tonnes of dust per year

During the last ice age when winds were stronger, there was much greater dust activity in desert regions than there is now. The huge amounts of dust blown out of the deserts in this period were deposited as loose windblown deposits in oceans and terrestrial basins. These sediments (some of which have a glacial rather than an arid origin) are called loess and are widely distributed across the globe. Extensive loess blankets exist in China, Central Asia, Central and Western Europe, North America and Argentina. In China, thick loess, with an average depth of 80–100 metres (260–300 feet), covers more than 273,000 square kilometres (105,000 square miles) on the Shensi Plateau. As a windblown deposit, loess is very vulnerable to wind erosion if disturbed, and it can thus act as a vast new source for dust storm activity.

Whirling dust

Dust devils are a common afternoon sight in desert areas. Termed "*djinns*" in Arabic, and "willy-willies" in Australia, they are high-velocity wind vortices made visible by the dust they carry. The low pressures in dust devils can create velocities approaching 22 metres per second (73 feet per second) at their centre. They are usually around 75–100 metres (250–300 feet) high, although exceptionally they may reach almost 1 kilometre (3,300 feet) high. They are short-lived phenomena and wander randomly over the desert surface. They are most frequently seen in river valleys and areas where dust accumulations occur.

Below *A duststorm in South Australia. Airborne dust particles are insubstantially small, but* en masse *they are dangerous and terrifying. Sweeping in from the desert, these rolling clouds of dust will paralyse transport systems, create countless medical emergencies, and leave residents facing weeks of cleaning-up and repairing damage.*

Bottom *A duststorm obscures a distant excarpment in Shaiaib Aswat in central Saudi Arabia. The dust comes from a number of sources, including chips broken from sand grains as they bounce along, old lake beds, and recent silt deposited by occasional floods. Dust storms in this region happen at all times of year, but vary in number and intensity with the season.*

MOVING SAND AND DUSTSTORMS

The wind has more effect where the ground surface is dry and free from vegetation. Serious wind-related hazards are thus common in arid and semi-arid areas, where both these conditions often apply. Duststorms, migrating dunes and blowing sand can all be dangerous.

Migrating dunes can present a major threat to agriculture, forestry, roads, railways and pipelines that happen to get in their way. The most serious problems occur around towns and villages where the surface is more disturbed and the pressure on land is at its greatest. Funnelling of sand along streets and high turbulence around buildings abrades building stone and can make living conditions very uncomfortable. In the sheltered zones in the lee of buildings, sand may accumulate and block roads and pathways. Communication lines are particularly vulnerable to dune encroachment – the blocking of roads often requires detours or costly removal measures.

Dune stabilization efforts have intensified in the Middle East during the past 30 or 40 years as a result of economic growth and development in the sandy areas. As well as the physical removal of dunes by bulldozers and dump trucks, many other alternatives have been tested. Sand fences constructed upwind of areas requiring protection have proven quite successful, but the fences are eventually buried by the sand they trap, so further fences must be built on top of the original ones. Fences must be carefully managed to prevent the creation of very large dunes which may themselves pose a serious encroachment hazard. The flattening of moving dunes is a temporary solution, but the flat surface soon becomes unstable and dunes redevelop. The most effective and permanent method of stabilization is to plant vegetation. However, in areas where water is scarce, vegetation cover is often difficult to sustain.

Right *A mud house that has been blasted away by sandy winds and then buried by migrating sand dunes on the outskirts of the Salah Oasis in central Algeria, North Africa. Many oases are surrounded by abandoned houses – or even entire settlements – like these that have resulted from booms and then busts in water supply or the availability of labour.*

Far right *A duststorm in Mali, West Africa, reduces visibility to a few metres (yards). Small duststorms, known as haboob, often occur at the front of desert thunderstorms at any time during the year. Larger duststorms, called khamsin, are caused by strong, steady winds blowing over the desert surface, and usually occur at certain seasons of the year.*

Below *Villagers clearing wind-blown Saharan sand from a house in Mauritania, West Africa. Moving dunes are most problematic around desert margins. In these regions, overgrazing, wood cutting and intensive agriculture can remove the scant natural vegetation that helps to contain drifting sand.*

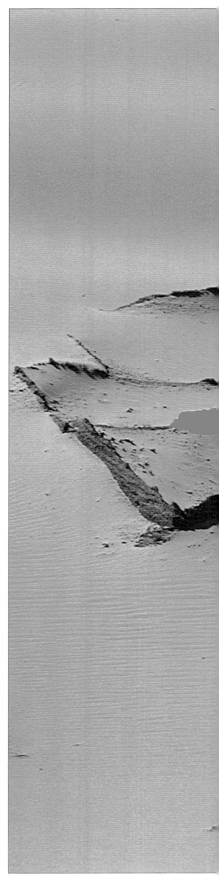

Airborne dust

Human activity in semi-arid areas provides the perfect environment for dust generation. In the Sahel of Mauritania, for example, many duststorms have occurred as a result of the agricultural activity on marginal land. The use of off-road vehicles in the Mojave Desert has caused duststorms extending for over 30 kilometres (20 miles) and covering 300 square kilometres (115 square miles). Surface disturbance caused by construction work and military manoeuvres – such as the North African campaign in the 1940s and the more recent Gulf War – can also initiate duststorms. Over-irrigation around the Aral Sea in Kazakhstan has reduced the water level drastically and has led to damaging duststorms of highly erosive saline dust of fine silt from the dry sea bed.

The damage inflicted by wind and dust is of many kinds. Crops may be damaged by sand-blasting, uprooting or defoliation by strong winds; cattle are often suffocated; and plant and human diseases are spread. Bronchitis, emphysema and conjunctivitis can all be aggravated by dust. "Valley fever" in Arizona, for example, is an airborne fungus that accounts for about 30 deaths a year. Severe reduction in visibility caused by dust also has an effect on human activity. Duststorms over the Arabian Gulf frequently disrupt movements of aircraft and disrupt radio communications. Dust activity on one major road in Arizona has become so frequent that a duststorm alert system has been installed. Automated road signs warn drivers of poor visibility ahead on the road and the frequency of accidents has been reduced.

This kind of hazard appears to be getting more serious as the pressure of human activity on marginal lands becomes more intense. The difficulties posed by blowing sand can be, to some extent, overcome by building design, and several techniques of stabilization can be applied to moving dunes. However, in some cases the hazards can only be surmounted by avoiding the affected areas entirely. Once dust is in the atmosphere it is very difficult to control. Dust hazards are therefore best tackled at their source by preventing surface disruption caused by human activity.

FLOODS

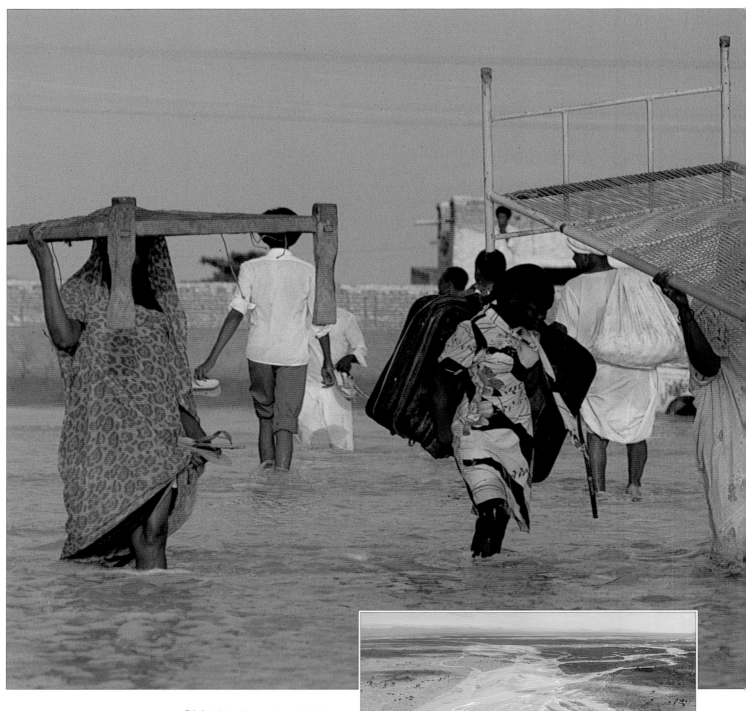

Right A shallow wadi in the United Arab Emirates is awash with floodwater after the heaviest rainfall in more than 10 years. Moisture from such events is quickly lost through evaporation and soakaway. Vehicle tracks have avoided the wadi bed, even when it was dry, because the alluvial sands deposited during previous inundations makes for difficult progress compared with the firmer ground on the sloping wadi sides.

Although deserts have little rain on average, on the occasions when rain comes the rate at which it falls can be dramatic. It has been said that in deserts more people have drowned in rivers than have died of thirst.

Arid regions experience the most unpredictable rainfall patterns in the world. In general, the less it rains on average in a particular region the less predictable rainfall there is. Both Bakel and Dakar in Senegal, for example, lie on the same 500-millimetre (20-inch) isohyet (line joining areas with the same average annual rainfall), yet a difference of 270 millimetres (11 inches) was recorded in 1972. Rain may fall only once every 8 years in parts of the Sahara and the Middle East, and only once every 18 years in desert regions of Peru.

The reason for the unpredictability lies with local climates. Most deserts are located beneath semi-permanent anticyclones (areas of high pressure) into which moist air, rain-bearing frontal systems or tropical cyclones can only occasionally penetrate. When they do penetrate, rapidly rising hot desert air, which cools and condenses as it rises, causes highly localized rainfall. One consequence of such localized rainfall is that surface runoff may flow into only a few *wadis* (occasionally flowing desert watercourses), which leads to torrents of water in a relatively small area.

Flash floods

In areas with a fair amount of vegetation and deep soil, most water generated by storms can be absorbed. In arid lands with only a thin covering of soil, however, the surface layers become saturated much more quickly and flow over the surface begins sooner. If the intensity of the rain exceeds the rate at which the soil can absorb the water, runoff rapidly courses down hills and fills channels to cause flash floods.

Heavy storms, while not very common, can devastate areas with steep slopes and thin soil. In Colorado Springs in the United States, for example, a flood of some 12 million cubic metres (420 million cubic feet) of water passed through the city in just two hours, causing immense damage. The stream bed, dry for most of the year, could have accommodated this amount of water if it had taken 20 hours or longer to fall, but the speed and intensity of the flood – rather than the total amount of water – created severe problems.

Flooding can also be a problem with permanently flowing rivers. The Nile, for example, breached its banks in 1988, causing much destruction in Sudan. In 1990, it was reported from Mogadishu that tens of thousands of people had been evacuated in southern Somalia when their villages, which are built in normally arid areas, were submerged under the flood waters of the Juba River. Such events are difficult to monitor and predict, and they do not benefit anyone because much of the water runs off into oceans and is lost.

Flash floods carry great loads of sediment. The reason is that soil accumulates on the surface between the rare floods and is then suddenly washed away all at once. In 1973, for example, on the Wadi Medjerdah in Tunisia, a storm – something that occurs there on average only once every 200 years – produced a deposit of silt over an area of about 138 square kilometres (53 square miles).

Above On the outskirts of Khartoum, Sudan, residents carry their furniture away from houses affected by flooding Khartoum is situated at the confluence of the White Nile and the Blue Nile, which thereafter form the main Nile River A sudden increase in the level of either tributary can cause the river to burst its banks, inundating low-lying areas of the Nile valley.

Most desert floods occur away from the isolated human settlements, and threaten only the unwary traveller. Centuries of experience have taught the Bedouin not to pitch their tents in wadis. Permanent rivers, which attract human populations, present a special problem. Controlling the floodwaters of the Nile has preoccupied local inhabitants for more than 5,000 years.

41

LIFE IN THE DESERT

In some senses, deserts are barren places, being sparse in plant and animal life, but there is an amazing variety of species well adapted to such harsh environments. Millions of years of evolution have produced some extraordinary ways of coping with extreme conditions of aridity and temperature. These range from cacti that consist largely of an immensely swollen stem to store and conserve water, to mammals with specially adapted water-conserving kidneys. Animals from tiny insects to large mammals inhabit deserts, including representatives of most major groups – even amphibians. Yet the very specialization of such flora and fauna makes them vulnerable to perturbations of their environments.

Above *The bat-eared fox is well adapted to desert conditions.*
Left *Desert blooms provide a rare blaze of colour in a California State Park.*

PLANT SURVIVAL

Life in a desert environment involves many problems for plants. Intense light can damage pigments and high temperatures can disrupt the biochemistry of cells, but the most serious problem of all is desiccation. Water is usually in short supply and adaptations that assist in water conservation are universal among desert plants.

The main problem for plants is that photosynthesis (the production of fundamental sugars using light energy) requires carbon dioxide (CO_2). To ensure efficient uptake of CO_2, the leaves of plants have tiny pores (stomata). But these also allow the evaporation and diffusion of water vapour from plants, a process that occurs rapidly in deserts.

Conserving water

Natural selection in the desert favours plant forms that have reduced surface areas through which water may be lost. Some cacti have adopted an extreme form. They have neither leaves nor stems and are almost spherical, providing the least possible surface area for a given volume. The number of stomatal pores is also reduced. One effect of this is to reduce CO_2 intake, so growth rates in these plants is inevitably very slow. Other species of plants avoid excessive exposure to the Sun and drying winds by growing largely underground.

Above *A variety of cacti and shrubs growing in the desert of Arizona, USA. Where conditions are suitable, succulents and shrubs can form a substantial and varied vegetation. In the foreground is an undergrowth of cacti and thorn bushes. Silhouetted against the sky are the "emergents" of the cactus forest, specimens of the organ-pipe cactus (Lemaireocerus spp.)*

Right *Quiver trees, pictured here growing on the fringes of the Namib Desert in South-western Africa, show a variety of adaptations to the arid climate. The trunk is short and bulbous and acts as a water reservoir; and the pale-coloured outer bark reflects sunlight. The cactus-like leaves, which have a low surface to volume ratio in order to conserve water, are confined to the ends of the branches, where the cooling of the wind is greatest.*

Left *Sodom's apple* (Caloptropis procera), *a member of the milkweed family, growing along the base of a sand dune in the Saudi Arabian desert. The forward edge of a dune is one of the most favourable habitats in a sand desert because of the availability of moisture. Rain water drains down through the dune and seeps out at the base.*

Above *A cactus flowering in the Mexican Desert. Cactus flowers typically have a waxy appearance and rarely last longer than a day or so. Some cacti flower only at night, and the blooms shrivel next morning. Cactus flowers are succeeded by fruits which may be smooth, scaly or spiny (for example, the prickly pear.)*

The root systems of desert plants are often extremely deep, tapping supplies of water many metres (feet) beneath the surface to replenish their own lost water. Among such plants are the melons and squashes (family Cucurbitaceae). But a surprising number of desert plants do not have deep root systems. The cacti, such as *Opuntia*, for example, have shallow, fibrous roots that are largely confined to the upper few centimetres (inches) of soil. These allow the cacti to catch water from dew formation or occasional showers.

Perhaps the most remarkable adaptation for a desert plant is the complete absence of roots found in *Tillandsia latifolia* from the Atacama Desert. It has stiff, spiny leaves arranged in a star-like fashion, but has no roots. The plant forms a ball that rolls across the desert, blown by the wind. When conditions are moist, it absorbs moisture from the air.

Protection from the Sun

Exposed to high-intensity sunlight, desert plants risk having their photosynthetic pigment (chlorophyll) damaged, and consequently use various devices to combat this. Some cacti, for example, are covered in long white hairs that reflect much of the incoming radiation. Others produce bulbous cells on their surfaces that shield the underlying photosynthetic cells and prevent the tissues becoming overheated. Air temperatures can exceed 55°C (131°F) during the day, and leaf surfaces may become even hotter. At these high temperatures, vital biochemicals – such as the enzymes (proteins) that control many cell processes, including photosynthesis – may be damaged and rendered inactive. Only those plants whose proteins can cope with such extremes of heat can survive. One of the most tolerant plants is *Tidestromia oblongifolia* which actually achieves its highest growth rate at temperatures of about 45°C (113°F).

In general, desert plants have a lower density of stomata in their leaves than plants from other habitats. This helps them to conserve water. But one desert plant, *Welwitschia mirabilis*, from the central deserts of Namibia, has a very high density of pores. The bulk of the water supply in this region comes in the form of mist and dew. The high pore density of *Welwitschia* allows it to collect this valuable moisture.

The problem of balancing CO_2 uptake with water loss has been partially overcome by succulents that open their pores only at night, when temperatures are lower and so less water is lost through evaporation. However, the light needed for photosynthesis is not available at night. Instead, the CO_2 absorbed is stored in a temporary form as an organic acid and then used during the day. This is called Crassulacean Acid Metabolism (CAM) after the succulent family Crassulaceae in which the process was first described.

Many of the desert grasses have developed an efficient technique for absorbing CO_2. Like CAM plants, they also fix carbon dioxide into an organic acid, but they then move it to specialized cells deep in their leaves, where it is stored and then used when it is needed for photosynthesis. This method ensures that, for a given amount of growth, these plants need open their stomata only briefly.

SMALLER PLANTS

The shrubs and succulent flowering plants are the most apparent form of vegetation to any desert traveller, but many other types of plant that play an important part in the ecology of deserts may not be so immediately evident. Among these are algae, lichens, fungi and short-lived annual and ephemeral flowering plants that spend much of the year hidden from view in one way or another.

Deserts might not appear to be promising habitats for mosses because they are so dry. Still, some species of moss, like those of the genus *Tortula*, are capable of survival, especially on rocky slopes and in crevices, where they receive some shelter from the direct sun. Mosses survive in the hot, dry conditions simply by drying out completely and becoming dormant until they are next provided with water. This often involves waiting an entire season and sometimes even several years. Such is the capacity of some of these plants to endure prolonged drought that dry specimens in museums have been known to recover and grow after 250 years without water. In addition, they can withstand temperatures as high as 55°C (131°F) while in a dry state.

Desert lichens

Even less conspicuous than the mosses, but far more widespread in deserts, are the lichens. Lichens are combinations of fungi with algae, or sometimes with blue-green bacteria (cyanobacteria). The algal cells live within the fungi, surrounded by a protective layer of fungal threads – called mycelia – which shade them from intense radiation and from desiccation. In return for this protective function, the fungus derives nutrition from the photosynthetic algae. The two live in mutually beneficial accord (symbiosis). An extra benefit for the fungus when the photosynthetic member is a cyanobacterium is nitrogen fixation: blue-green bacteria are able to convert nitrogen gas in the atmosphere into useable nitrogenous compounds. This can prove beneficial to the whole ecosystem. Lichens have no root systems, but absorb water vapour from the atmosphere.

Lichens occur in a range of different forms. Some are projecting, leaf-like structures, others branching mats, and some superficial crusts on rocks. These "crustose" lichens are extremely common on rocks in the desert wherever surface conditions are reasonably stable, but they are easily overlooked. Even less conspicuous than these superficial (epilithic) lichens are the lichens that live beneath a surface layer of rock (endolithic lichens), inhabiting the minute pores and crevices within the rock, especially in limestones. The living lichen occupies a layer up to 1 centimetre (0.4 inches) deep in the rock, where light still penetrates but where the delicate algal cells are protected from drying out. Probably the greatest problem that these rock inhabiting lichens face is that the carbon dioxide they need from the air for photosynthesis reaches them only slowly.

Like the mosses, lichens become dry and dormant during unfavourable conditions and in this state they are extremely tolerant of both high and low temperature. The lichen *Ramalina maciformis*, for example, is not killed until the temperature reaches 85°C (185°F).

Short lives, long lives

Some desert plants are active only for limited periods. These are known as the annuals and ephemerals, and included among them are poppies, rockroses, grasses and chenopods. These plants spend most of the year (and most of their lives) as dormant seeds, only germinating under favourable conditions – when rain falls, for example. They then complete their life cycles very briefly, with a quick burst of flowering. Annuals and ephemerals devote most of their productivity to the formation of new seeds that will lie in dry soil until the next moist period. Mud bricks from archaeological sites have been found to contain seeds that can still germinate after 300 years of interment. The seeds, also, sometimes have to survive strong concentrations of salt.

Many of these plants have evolved special dispersal techniques to ensure that their seeds have the best possible chance to germinate. Some desert grasses, for example, produce a large number of seeds that become entangled, forming a ball shape that is easily blown through the desert. The individual seeds have sharp points that catch in the surface, detaching the seed from the ball. Eventually, the ball disintegrates, the seeds having been successfully dispersed. Other plants produce seeds with hooks, or barbs, that become entangled in the fur of grazing animals' legs.

Left *Lichen growing on a granite outcrop in the Arizona Desert, USA. Lichens are among the hardiest of all life-forms, and are found from the coldest polar regions to the hottest deserts. These lichens are thriving in conditions that have killed the desiccated agave plant visible at the top of the picture.*

Below *A rare sight,* Calandria discolor *flowering in the Atacama Desert of South America. For many desert plants, flowering is an expensive necessity that uses up vital resources. Most flowering desert plants bloom only very briefly, and reproduction becomes a race against time. Fertilization must occur before the exposed reproductive organs are damaged by the intense heat and light.*

Above *"Living stones", or lithops, growing among pebbles in the Namib Desert of southwestern Africa. These plants grow partly underground, exposing only a small photosynthetic surface that resembles a rounded stone. Each species (*Lipidaria margaretea *is pictured) has its own distinctive shape and colour. As well as serving to conceal the plant from herbivores, the coloration also minimizes the temperature differential between the plant and its surroundings. Lithops tend to occur in open areas where the wind keeps the surface free from sand.*

DESERT INVERTEBRATES

Size is an important feature of desert invertebrates because the main route of water loss is through their surface. The smaller the organism, the greater its surface area relative to its volume. Not only do small organisms have relatively more surface area than large ones, but they also have a smaller volume of fluid to start with. Water loss, therefore, becomes increasingly critical for smaller species, and many adaptations to the problem are found in the invertebrates.

Following heavy rains in the desert, pools often form, providing a temporary source of water. These pools contain a surprising number of protozoa – very common, single-celled aquatic animals, including amoebae and ciliates. As the pools dwindle these protozoa secrete a tough coating, or cyst, and survive in the soil in a dormant state until rain returns.

These temporary pools frequently contain many crustaceans, including familiar marine invertebrates such as shrimps, lobsters and crabs. Within a few days of their formation, desert pools can contain populations of small planktonic water fleas and cyclops. Within a week, the phyllopod *Triops* may be found. These animals grow to 3 centimetres (1.2 inches) in length and compete with tadpoles for the detritus in the water. *Triops* lays resistant eggs before dying as the pools dry up.

Desert insects
Although they are small, insects have solved the water loss problem by having a special waxy, waterproof layer to their cuticle (protective outer layer). Consequently, insects are well able to cope with desert life and are well represented in deserts, without having many special adaptations.

Desert insects fall into two main groups. Many species have the mobility of flight, which allows them to escape from the harshest conditions. After the rains, they can return to make use of the desert's resources. The other insect group contains those species that are poor fliers or have no ability to fly at all. These are permanent residents.

The desert locust (*Schistocerca gregaria*) is a powerful flier and migrates over vast areas of northern Africa and the Middle East. In common with other locust species, *Schistocerca* periodically forms huge swarms which devastate vegetation, causing catastrophic losses to crops.

Locusts are grasshoppers that herd together and move coherently. They exist in two forms, called solitary and gregarious. The solitary form of the desert locust looks and behaves quite differently from the gregarious form. What induces the change from one form to another is not fully understood, but is thought to involve pheromones (airborne chemical messengers). If solitary individuals are crowded together they change to the gregarious form. Crowding occurs particularly when winds and rains converge. Distinctive winds associated with the monsoon bring the locusts together, and rains provide new plant growth to feed on and moist soil for egg laying. The population builds up locally and the switch from one type to the other takes place. The gregarious phase may then persist for several years until strong winds and storms disperse the swarm so that the isolated individuals revert to the solitary form.

Beetles are well suited to desert life. Scarab beetles (family Scarabaeidae) were revered by the ancient Egyptians, and were thought to represent life. Scarabs dispose of the dung of camels, goats and donkeys by breaking off small portions which they roll into balls and bury, with an egg laid in each ball. Dung may also be buried and used as a food store for adults. The tenebrionid beetle *Onymacris unguicularis* makes use of night-time fog common in the Namib Desert. It adopts a curious "head down, bottom up" posture on the dune crests so that when the fog condenses on its body the water runs straight down to its mouth.

There are a number of predatory species. Centipedes, up to 20 centimetres (8 inches) long, use poisonous claws situated just behind their mouths to catch and kill prey. Scorpions' normal food consists of insects and spiders, which they grab with their large, crab-like chelae (pincers). They kill with an injection of venom from the sting at the end of their tail. Some of these venoms are potentially fatal, even to people. The solifugids, or camel spiders, are especially characteristic of African deserts. They have a pair of strong, vertically operating chelicerae (pincers) in front of the mouth, which are used to catch and hold prey. They eat other invertebrates, lizards, mice and even small birds.

Above *A species of darkling or tenebrionid beetle (*Onymacris plana*) drinking condensed fog in the Namib Desert of southern Africa. Another Namib tenebrionid condenses the fog on its own body. Although their impervious cuticle reduces evaporation to a minimum, all desert insects inevitably lose water through respiration. Insects such as the tenebrionids have adapted to make use of even the most transient of water sources. The Namib receives most of its moisture in the form of early morning sea-fogs which roll in from the coast. The moisture in the fog condenses on surfaces that have been cooled by the desert night, and is briefly available for drinking.*

Below *A desert scorpion stands exposed to the Sun on a gravel plain in the Namib. Scorpions spend most of the day in burrows or beneath rocks, sheltering from the Sun's heat, and only venture out when the shadows lengthen. Desert scorpions are found with both dark and light coloration, because there is no distinct advantage to being one or the other. The advantages and disadvantages of both options (in terms of camouflage, heat loss and absorption) are about equal.*

Above *A female Australian desert wolf spider (*Lycosa storri*) carries her egg sac attached to her abdomen as a safeguard against desiccation. Spiders are the commonest of the desert's invertebrate predators, and many families are represented. However, the relative scarcity of flying insects means that web-spinners are proportionately few in number. The majority of desert spiders are either active hunters or species that ambush their prey.*

MAMMALS

In the harsh desert environment, animals face three major, interrelated problems: high temperatures, shortage of food and, above all, lack of water. Although deserts are such harsh regions, they are home to a surprisingly wide range of animal species, including many mammals. Large desert mammals include species like camels, donkeys and goats, which have all been domesticated. There are also gazelles and the Arabian oryx (*Oryx leucoryx*), a threatened species. In the Namib Desert of coastal southern Africa, elephants and giraffes can also be found. In Australia things are quite different. The placental mammals just described are absent, although the kangaroos, which are marsupials, have similar lifestyles to such animals as gazelles. Animals that are not native species, such as camels, have in fact been introduced to the Australian desert. These are all herbivores, feeding on the sparse desert vegetation. Large predatory mammals are usually absent from deserts, although the Namib and Kalahari are exceptional in having some lions and jackals.

Smaller mammals include various ground squirrels, kangaroo rats, jerboas, gerbils and pocket mice. There are small carnivores, including the fennec fox (*Fennecus zerda*) in Africa and the kit fox (*Vulpes velox*) in North America. In Australia there is the mulgara (*Dasycercus cristicauda*).

Coping with the heat

Animals of different sizes cope with high temperatures in different ways. Large mammals cannot escape the Sun's heat as there is virtually no shade. Most have a thick coating of hair on their upper surface which insulates the tissues beneath and keeps them relatively cool. The outer hair can reach very high temperatures: 70°C (158°F) has been recorded on the outside of a camel while the body temperature remained around 40°C (104°F). A light-coloured coat helps to reflect some of the heat of the Sun. While an animal's upper surface is well insulated, the undersurface, which is in shade, is often almost bare so that heat can be easily lost.

Physiological adaptations allow many large mammals to tolerate an increase in body temperature. Mammals and birds are homoiotherms – that is they normally keep their body temperature constant. But some animals can adapt to heat by being able to tolerate a temperature rise. These animals can save on water that would otherwise be lost through sweating or panting. For instance, the body temperature of a camel may change more than 22°C (40°F) in a single day. During the cold night, the camel allows its body temperature to fall well below normal, then during the day the Sun warms the camel, past normal to a high of 41°C (106°F).

Above *A North American blacktail jackrabbit (Lepus califoricus) shelters from the strong desert sunlight. Rabbits are temperate animals, but they have colonized deserts throughout the world. Their burrowing lifestyle is well suited to the desert environment. The animal's large ears contain a dense network of blood-vessels, which radiate away excess heat. Although expensive in energy, the rabbit's rapid, jumping style of locomotion minimizes the animal's contact with the hot, Sun-baked desert surface and so reduces heat absorption.*

Right *A herd of South African eland (Taurotragus oryx) at speed. Eland and other antelopes have an arrangement of blood vessels in the head and neck that keep the brain several degrees cooler than the rest of the body. Otherwise, strenuous activity, such as running, would tend to raise the animals' brain temperatures to dangerous levels.*

Below *A group of meerkats (Suricata suricatta) warming themselves in the morning Sun of the Kalahari Desert in South Africa. Meerkats are social animals that live in communal burrows. Food collection on the surface is also a communal activity. The upright stance improves the animals' chances of spotting a predator.*

Small desert mammals deal with heat by avoiding it. They usually spend the daytime underground in burrows and emerge at night when it is cooler. Only a few species emerge briefly during the day, before returning to their burrows to cool off. The African ground squirrel (*Xerus erythropus*) uses its bushy tail as a parasol when it forages in the day. Desert rodents do not have sweat glands. In times of heat stress they dribble saliva onto their throat hair to cool themselves.

Saving water

With the exception of oases, there is little free drinking water available to desert mammals. However, given access to water, camels and donkeys can drink large amounts – between 20 and 25 per cent of their body weight in a few minutes. Large mammals dig holes in dry riverbeds to reach water sources below the surface. Many animals lick or suck off the dew and fog that condenses on plants.

Much of the water available to animals comes from their food. Small, seed-eating mammals store seeds in their burrows. As the animals breathe, they exhale water vapour which humidifies the air in the burrow. This moisture is absorbed by the seeds and ultimately returns to the mammal when the seeds are eaten. Carnivores and insectivores meet their water needs through the body fluids of their prey.

Water is important as a coolant, but it is also vital that the concentration of the body fluids remains constant; hence the mechanisms allowing temperatures to rise so reducing the need to sweat or pant. But water is also lost, along with dissolved waste products, in urine. The kidneys control the water content of an animal's body. In desert mammals, particularly smaller species, the region of the kidney where water is reabsorbed from the urine is enlarged. This allows them to produce a very concentrated urine.

ADAPTING TO EXTREMES

Almost unbelievably, in a habitat characterised by a virtual absence of any standing water, both fish and amphibians are found in the deserts. In one of the hottest deserts on Earth, Death Valley on the Nevada-California border, there are some small fresh or brackish (salty) water springs. Several species of desert pupfish (*Cyprinodon* spp.) live in these streams. The fish can tolerate temperatures as high as 43°C (111°F). On cold winter nights, these same fish also have to tolerate temperatures as low as 1 or 2°C (34 or 35°F). It is this considerable temperature tolerance, which is very unusual in aquatic animals, that has enabled the pupfish to survive in Death Valley for at least the last 30,000 years. These fish graze on blue green algae, which are also highly tolerant of large temperature changes. Today the greatest threat to these tiny, uniquely hardy fish is not the harshness of the desert but people. The growing demand for water in California is lowering the water table and some of the springs where the pupfish live are beginning to dry up.

Right *Because of its unusual sidewinding motion, in which the head and body are thrown forward at an angle to the direction of travel, Peringuey's viper (*Bitis peringueyi*) is often known as the "sidewinder". This small viper is found mainly in desert areas of South Africa and Namibia. This type of motion is an efficient method of crossing soft sand. It also allows the snake to reduce the amount of its body in contact with the hot surface of the sand, and so avoid overheating.*

Reptiles and amphibians

Fish need permanent water, but amphibians need water only for breeding. Consequently, they can make use of temporary pools that form during irregular rain storms. Amphibians normally lose water through their skins. However, desert species can tolerate a water loss of up to 50 per cent of body weight. Some species retain a dilute urine in a pouch of the digestive tract, rather than excreting it and losing the water.

Desert dwelling amphibians, such as the spadefoot toads (*Scaphiopus* spp.), pass the time between rains by aestivating. Aestivation is the hot-climate version of hibernation: the animal becomes dormant with a reduction in its energy needs. It is a technique used by many groups of desert animals to deal with periods of adverse conditions.

Among the mammals, ground squirrels aestivate, as do birds such as the poorwill (*Phalaenoptilus nuttallii*) and some reptiles. Often, aestivation occurs when food or water is scarce. Spadefoot toads burrow down almost one metre into

Above *A few desert lizards, such as the slow-moving yellow and black Gila Monster (*Heloderma suspectum*) of the southwestern United States, have toxic venoms to ensure the rapid death of prey and reduce the risk of injury during the capture. Venom is not only useful for catching food, but is also used for protection against predators. Like other desert animals, the Gila Monster can withstand long periods when food is unavailable.*

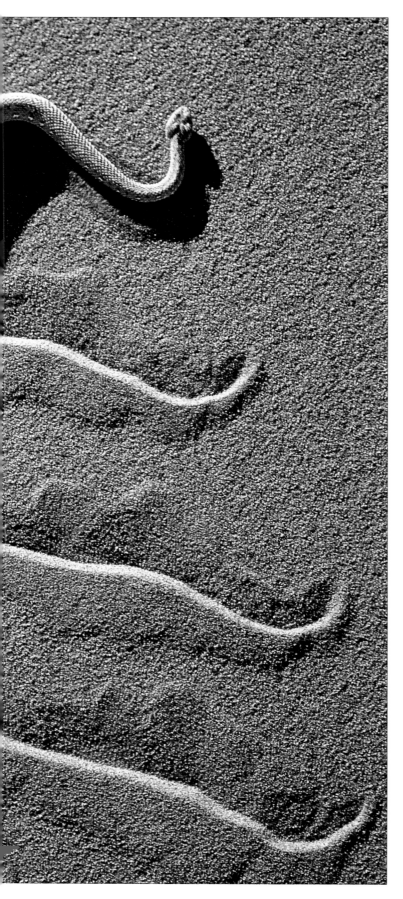

moist soil. Their outer skin layers then become harder and leathery to reduce water loss. In this situation, their metabolic rate drops dramatically, and they survive for up to 9 months on fat reserves. With the onset of rain, the amphibians become active again, burrowing back to the surface and shedding their skin. Mating and egglaying take place in freshly formed rainwater pools.

Compared to amphibians, reptiles are far more suited to desert life. Lizards, snakes and tortoises all have thick, water-retentive skins. Most desert lizards and tortoises are diurnal (active during the day), while many of the snakes are nocturnal. In the summer, lizards often avoid the hottest times of the day and are active in the morning and evening only, moving from light to shade to regulate temperature.

Desert tortoises (*Gopherus agassizii*) are herbivorous, as are a few lizards. Most lizards and all snakes are carnivorous. Insects are the main food for smaller species, but the larger lizards and snakes feed on other lizards, rodents and eggs.

Desert birds

Perhaps just because they can fly, birds are common in deserts. In terms of body size, most are small. Birds have higher body temperatures than mammals, and can tolerate as much as 45°C (113°F) for long periods of time. Most desert birds reduce activity during the heat of the day, since this will also elevate body temperature.

In the absence of trees, nests tend to be on the ground. Eggs can be buried by sand or become overheated if left uncovered for any period of time. In addition, there is always the possibility of the eggs being eaten.

The bird equivalent of the camel is the ostrich (*Struthio camelus*). This large flightless bird is often found in sizeable herds, mainly in southern Africa. Its fine feathers provide a thick insulation against the sun. By drooping the wings the sparsely feathered, thoracic region is shaded and can radiate excess heat. Like other birds, ostriches cannot sweat to keep cool. If heat-stressed, the ostrich pants, losing water vapour from its throat and so cools the body.

THE DESERT ECOSYSTEM

The desert ecosystem consists of a community of living organisms interacting with its non-living environment. The energy from the Sun fuels photosynthesis, and the organic matter of plants provides energy for herbivores, which are themselves preyed upon by other animals. Carbon dioxide gas, and nitrates from the soil are taken up by plants and made into living organic matter. As dead organisms are consumed by fungi and bacteria, many of their component molecules are released into the non-living world.

Energy input

The input of energy into the desert ecosystem comes from the photosynthetic production of new plant material. The rate of this "production" in deserts is generally low – very similar to the rate in arctic tundra. The growth in plant material is erratic and depends on water availability.

The production of new plant material may fail completely if there is prolonged drought, but the minimum amount of precipitation needed for plant growth varies. The North American hot desert ecosystems, for example, need 38 millimetres (1.5 inches) of annual precipitation to maintain the growth of perennial plants, but in the Namib Desert an annual grass, *Stipagrostis gonatostachys*, is able to grow when precipitation is as low as 10 millimetres (0.4 inch) per year.

The energy fixed from sunlight by plants becomes available to grazing animals, ranging from grasshoppers, bugs and aphids to gerbils, ostriches and oryx. Not all the energy finds its way into the consumers, however. Some is used by the plants, or is lost when plants die.

Food webs

The feeding relationships of animals in the desert, as in other ecosystems, are very complex and form a food web in which different levels of feeding ("trophic levels") can be distinguished. Animals may occupy several different positions, depending on what food is available. Because the passage of energy from one trophic level to the next is relatively inefficient, especially in deserts, there are usually smaller numbers of animals (and lower biomass – the total mass of living things) at each succeeding level. Within the ecosystem, interactions between different components often lead to predictable, cyclic processes. A wet period may lead to a burst of vegetative production and an explosion in the population of invertebrate grazers. This may be followed by a build-up of predators and scavengers.

Vegetation and the environment

Vegetation has a considerable effect on the physical environment, creating patches of shade, reducing windspeeds and protecting the soil surface from the effects of rain impaction and surface runoff. Even a lichen crust has an important role in surface stabilization, because it protects the soil from the rain and absorbs much of the moisture, reducing surface runoff. The vegetation cover as a whole reflects the Sun's rays and creates a thermal blanket, which prevents the surface becoming excessively hot and so modifies the movement of water in the soil. The removal of plant cover by overgrazing and fuel-wood collection can, for these reasons, lead to desertification.

Right *An elf owl (*Micrathene whitneyi*) shelters in a giant saguaro cactus (subfamily Cereiiodeae) in Arizona, USA. Cacti are the trees of the desert in more ways than appearance. This saguaro provides a micro-habitat that supports a wide variety of animal life in addition to the watchful inhabitant in the upper storey. In the basement, mice, lizards and invertebrates all burrow into the relatively moist ground to be found near the base of the plant.*

Below *A herd of giraffes drinking from a freshwater spring at the edge of the Etosha saltpan in East Africa. Waterholes are the most favourable habitats in the whole of the desert ecosystem and attract a wide variety of visitors over great distances. However, the scarcity of food resources means that permanent residents are few.*

PEOPLES OF
THE ARID WORLD

The variety of human societies in deserts is as great as the variety of deserts themselves. Yet all these peoples share certain needs related to their harsh surroundings: the need for protection from extremes of heat and cold; the need for shelter from the raging winds that often characterize desert regions; the need to spend a large amount of time searching for food and water. The ways of life of many desert peoples are under threat. Damage to desert environments upsets the delicate balance upon which traditional livelihoods depend. Economic disruptions arising from population growth and periods of drought threaten people's lives. But there is still as great a richness and diversity of cultures as can be found in almost any environment on Earth.

Above *Camels are basic to the Tuareg lifestyle.*
Right *A campfire provides warmth as night falls in India's Thar Desert.*

DESERT PEOPLES

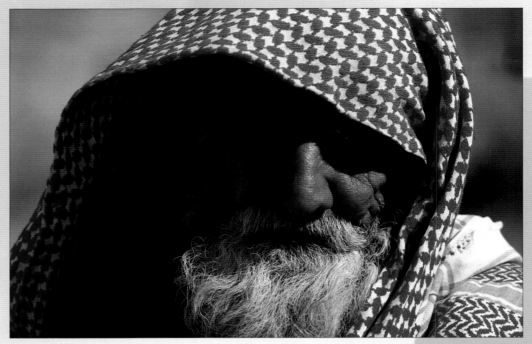

Left *A Saudi Bedouin, one of the dwindling number of nomads who traditionally evaded control by retreating to the desert fastnesses. In the last thirty or forty years, most bedouin have been forced to give up their wandering ways, although a few still maintain their traditional lifestyle in the desert.*

Below *A woman of the San people in Botswana, Africa. The San, together with the Khoi-khoin, form the Khoisan group, whose homeland is in southern Africa. Like the Australian Aborigines, the Khoisan have an ancient desert culture, which is carefully adapted to their hostile environment.*

The earliest known remains and artefacts of what are considered to be true humans have been found in desert regions, particularly those of East Africa. These regions were not, however, deserts at the times from which these finds date. Nonetheless, some of the earliest human cultures did exist in these same regions after they had become desert.

Desert peoples, particularly those of the Nile, the Tigris-Euphrates and the Indus river valleys, were also responsible for creating the first agricultural economies, characterized by the development of settled, irrigated farming. Similar economic revolutions took place in the high deserts of South America and in the southwestern United States.

Culture and religion

The core of the world's most extensive desert region, the Sahara and the neighbouring parts of Asia, is closely associated with the Semitic peoples, particularly the Arabs. They now are the main group living in a majority in countries of the Middle East. Other ethnic groups, however, do exist in the region. In the North African countries of Morocco, Algeria, Tunisia and Libya, for example, there are significant Berber populations. On the southern side of the Sahara most peoples are of African origin, but over the last 1,000 years their distinctive character has been modified by long-standing connections with the peoples of the northern Sahara. The deserts of southern Africa have become a refuge for the traditional peoples of the Kalahari and the Namib deserts.

The majority of the peoples of the arid regions of the Central Asian republics, as well as the dry areas of Turkey, are Turkic in origin and culture. In the Gobi Desert, most people are Khalkha Mongols. The geographical positions of these deserts have, however, has meant that their populations have often been exposed to influences and migrations from many surrounding areas.

Outside the Old World, many desert areas were settled by Europeans during the nineteenth century and the first part of the twentieth century, usually displacing or subjecting the original population. In North America, the original desert peoples of the Southwest, such as the Navajo and Apache, were deprived of their land by European intruders, dispersed or relocated, and their numbers severely reduced. Since that time, there has been a massive migration of settlers into the Sun Belt states from the multi-ethnic population of the contemporary United States. Large numbers of the descendants of the original Mexican people and those of mixed Mexican-Spanish origin live in the deserts of Mexico.

The mountain deserts of the Andes of South America are inhabited by the descendants of the Indian peoples who civilized the region during the past 1,000 years. The Atacama Desert of South America is such an extreme environment that the populations there have always been small; significant settlement has only been recorded in the modern era.

The Aborigines of arid Australia were severely reduced in numbers during the European invasion of the 1800s onwards. Some were displaced from their homes and cultural sites to make way for mineral exploitation.

Modern deserts

All the deserts of the world have much higher populations today than at any time in the past. This is particularly true of the deserts of North America and Australia, where major urban centres and some large cities – Phoenix, Tucson, Salt Lake City and even Los Angeles – exist in arid regions. Mining and agricultural development attracted settlement in a few of the dry tracts of Australia, while in the Middle East, oil exploration and subsequent development have caused substantial urban and industrial centres to blossom. The origins of the people constructing and then living in these new population centres are often very diverse. One example is that there are millions of people from the south and southeast of Asia living in the Gulf region.

Today's deserts contain a great deal of evidence of traditional societies and cultures, but almost everywhere the lifestyles of their populations are closely integrated into the modern world. In addition, because many economic activities in the world's deserts provide high mineral- and oil-derived revenues, the international community takes a great deal of interest in their internal affairs; a consequence is that the communities who reside there are often very international.

Above *Muslims at prayer in the Khajezud Murrad Mosque, Samarkand, Uzbekistan. The desert peoples throughout most of the Old World came under the influence of a strong unifying culture with the rapid spread of Islam during the 1600s and 1700s. Islam has remained a powerful influence to this day and is, indeed, increasing in global importance.*

Right *A nomad woman in Burkina Faso, Africa, bedecked in traditional finery. Like the Arabian Bedouin, the nomads of the Sahel are dependent on their domesticated animals. However, while the Bedouin lifestyle revolves around the camel, the African nomads depend upon herds of cattle, which they use, like the camel, as a source of meat, milk, hides and transport.*

SHELTER

Dry deserts with little vegetation place great demands on the ingenuity of the peoples who live in them. In hot deserts, the need is to insulate people – and where possible livestock – from the extreme midday heat. It is also necessary to ensure that living spaces are ventilated and that air circulation and cooling breezes are as effective as possible. There is an additional challenge in areas with cold nights. In cold deserts, the main need is still insulation, but from the cold, and especially the extreme cold of winter in some areas.

Building materials

Materials for construction are scarce in both hot and cold desert. As a result, some of the earliest dwellings of all were caves, and there is widespread evidence of this from the Mesolithic and Neolithic periods in the Sahara and in the Middle Eastern deserts. In northern Africa, natural caverns were enlarged and in a few places were used until the very recent past. Similar dwellings were created in the dry areas of what is now the southwestern United States.

Materials for the construction of permanent and temporary shelter have, throughout human settlement of deserts, come from the ground, from plants and from animals. Deserts are rich in rock cover and in some places also in clay deposits. Hides, wool and animal hair are relatively abundant in relation to the small human populations of deserts, although supplies of useful vegetation are often poor. Permanent dwellings traditionally were built with mud brick and brick in areas where clay occurs. River valleys have provided readily accessible material for the manufacture of sun-baked mud

bricks, and later for kiln-fired bricks. During the major ancient civilizations that graced the deserts of the world, the excavation and the transport of massive quantities of ambitiously engineered and skilfully dressed stone were important industries as towns and cities grew.

There have been urban centres in the deserts of the world from the earliest times, but the deep deserts can only be inhabited by people who adopt a mobile lifestyle that enables them to use the scarce fodder resources of the remote desert tracts, and these people need either mobile or temporary shelter. Tents made either from cloth or animal skins have been the main type of mobile shelter used by nomadic desert peoples for the past ten thousand years. The *ger*, a hut made of cloth stretched over a wooden frame, is widely used by the nomads of Central Asia. Temporary shelter can be constructed very easily where there are shrubs and herbs from which woody material can be collected.

Modern shelter

High technology has transformed the options available to modern desert dwellers. Artificial climates can be created in efficiently insulated structures with effective ventilation and air conditioning. The temporary nature of the traditional tent is emulated in the modern air-conditioned trailer. Such dwellings are not used by traditional livestock rearers, but largely for transient mineral and oil exploration camps. The numerous and rapidly developing permanent settlements of the oil-rich Gulf and of the Sun Belt of the United States have attracted many of the world's best architects.

Above Bedouin tents pitched above a broad wadi in Saudi Arabia. The low, spreading tents made of black cloth and supported by long guy ropes, are a traditional and highly effective solution to the problem of making shelter amid the heat and winds of the open desert.

Below Inside a tented classroom a teacher in a traditional Islamic school supervises an arithmetic lesson. Walls and roof of tightly woven matting exclude most of the light and heat, and create a cool, shady atmosphere. The desert sand provides seating for pupils.

Above Village women near Lake Dow in Botswana constructing a new roof for a circular mud-house. The thatched reeds will provide an indispensable rainproof "umbrella" to protect the Sun-baked mud walls. Without this protection, the house would literally dissolve during a prolonged rainstorm.

Inset left Camels laden with houses in northern Kenya. Their nomadic owners have travelled south from the Sudan. The curved saplings provide the framework and are covered with animal hides. Wooden blinds permit light and ventilation, and woven grass mats are placed on the floor, with a few sheepskins for comfort.

SHELTER: THE ISLAMIC TRADITION

In Islamic architecture, the style of construction of dwellings depends on several factors: the physical environment, including water supply, temperature, wind and rainfall; and cultural elements, such as social patterns and the strictures of religion. In addition, with town dwellings, the urban environment itself directs the development of individual house style and structure.

To illustrate the development of Islamic urban settlements and houses suitable for hot deserts and arid environments, we can take Iran as an example. There are some religious and social factors unique to Iran that have influenced the way in which houses there have developed. The elements of construction influenced by the requirements of climate and physical environment, however, are reflected in other desert regions of the Islamic world – and in some non-Islamic areas.

Urban development and the physical environment

In Iran, in a way typical of arid countries, cities have grown up largely in areas with an accessible water supply and sufficient agricultural land for the needs of the population. The main source of water in the region is from artificial subterranean channels, the qanats, which govern the distribution of urban settlements. The qanat system that surrounds the great northern salt desert of Iran, for example, is reflected by the urban developments that encircle the plain. Another factor is the winds that characterize the area and which curtail the use of the plains for agriculture or settlement. Political and strategic (military or other) demands also influence the position of urban areas.

Modern cities in Iran are increasingly built without taking climatic conditions into account. Such cities are made possible by the use of very large amounts of energy to drive the equipment needed to make them habitable, such as air conditioning systems and water pumps.

The design of houses

The usual pattern of a house is a central open courtyard surrounded on two or three sides by rooms. Houses are built in a cell structure, that is, sharing walls with neighbouring buildings, usually on three sides, leaving the fourth side facing to the street. The walls are normally thick, made with layers of clay or mud bricks, to restrict the conduction of heat into the house. The ground in the courtyard area is often somewhat below the level of the street to increase the amount of shade and to help retain cooler night air during the day and so reduce temperature inside the house. High walls around the courtyard add to this effect.

Traditional houses, particularly those in the richer quarters, have a pool or even a fountain in the courtyard, again serving to reduce the temperature inside the house. Also largely restricted to richer householders is the construction of a garden in the courtyard. A garden could be used to provide additional food, but its primary use, beyond the purely decorative, is once more to cool the house. Internal gardens are not only found in private buildings: mosques and other public structures, particularly those from the medieval period, often have garden courtyards.

Even in the poorer quarters of traditional Iranian cities, houses often include north-facing rooms that are designated for use during the hotter months. These rooms often hold another pool of water – in addition to the courtyard pool.

Unique to Iran, and of particular interest to the modern architect attempting to reduce reliance on artificial air-cooling systems, is the *bad-gir*, or wind-catcher. The *bad-gir* is a tower rising from the summer room and reaching to about 20 metres (60 feet) above ground. The tower has two sets of vents at the top: one smaller set facing the prevailing wind, and a larger set facing away. Relatively cool wind is channelled into the tower. The greater density of the cool air causes it to sink down into the house, while the less dense warm air inside the building rises to the rear vent and is expelled. The downward air is often cooled further by channelling it over a pool or through a filter of damp leaves.

Extremes of winter climate on the desert margins lead the more affluent householders to build winter quarters. These are on the south side of houses to benefit from the heat absorbed by the walls from the winter sun. As might be expected, doorways and air channels are smaller than in summer quarters, and the circulation of air is further restricted by textile coverings over doors and windows.

Above *The interior of the Shan Mosque, Isfahan, Iran. High, arched spaces, windowless walls and ceramic tiles all combine to produce a cool contrast with the desert outside. The blue coloration psychologically enhances the cooling effect, and many private dwellings also have their interior walls painted blue.*

Right *Multi-storey town houses at Shibam in southern Yemen are among the most spectacular examples of mud-brick architecture, and represent a thousand-year urban tradition. Buttressing of the lower floors spreads the load, and provides additional protection against the floodwaters of the walled-off wadi in the foreground.*

CLOTHING

The clothing of desert peoples depends largely on the temperatures of the desert where they live. The Australian Aboriginals, for example, originally wore only loin cloths, and sometimes dispensed even with these. Their naturally dark skin pigmentation is sufficient to protect them from the Sun, while temperatures at night are rarely low enough to warrant clothing. If temperatures did drop, Aborigines have been known to use fur blankets. The nomadic herdsmen of Tibet, on the other hand, wear robes with long sleeves and high collars, along with a sheepskin hat and perhaps a woollen or sheepskin wrap to protect them from the cold that prevails in their high-altitude environment.

Clothing must be durable, not only because of the scarcity of materials, but also because the crafts involved in creating materials for clothing are very labour intensive. Many of the desert dwellers of the Americas, such as the Pueblo, Navajo and Inca peoples, for example, took great pride in their handwoven garments. The Pueblo, in particular, were highly skilled at weaving and making cotton clothes, some of which were exceptionally colourful. They spun yarn from cotton and turned it into cloth. The men wore cotton breeches and kilts, and the women dressed in cotton wraps. Around their legs, both the men and women wore buckskin leggings. Later, when the Spanish introduced sheep, the Pueblo Indians also made woollen clothes. Other tribes of southwest America, particularly the Navajo and Apache peoples, originally wore skins from the animals they hunted, but their increased interaction with the Pueblo encouraged them to make cloth.

Clothing and culture

Among the cultures that used clothing, the designs were often closely associated with the religion of the area. In the vast desert regions where Islam originated, for example, similar basic designs occur over very extensive areas. In the desert regions where Muslim tradition has been a strong influence, the conventions of modesty and cleanliness prevail. Figurative display is absent and geometric Arabesque decoration is preferred on artifacts as well as clothing. The personal modesty required by the religion and culture of many of the desert people, especially those of Asia and the Middle East, reinforces the tendency to cover the body to protect it from the heat, and it is common in Muslim countries to cover the whole body, including the head.

Headgear is a distinguishing feature of male members of the nomadic groups of Arabia and northern Africa, and women also often wear distinguishing head-dresses. The veil for women is not, however, common among nomadic peoples, and women members of many such tribes participate freely and without constraint in the economy and daily life of their communities. Footwear was easily created from the skins of the livestock and sandals of simplicity and considerable elegance have long been and remain a part of the normal dress of the peoples of the desert, although going barefoot for long periods is also common. The desert surface can become uncomfortably hot to the tread after many hours and the protection of the sandal is often essential if distances have to be covered at hot periods.

Above *Market day. The clothing of desert peoples is usually fairly dark. Such colours have the best insulating properties and desert dust shows up least against earthen hues. However, given the opportunity, desert-dwellers display the same love of finery and adornment as the rest of the human race, as is shown by this group of women in traditional dress.*

Top right *Cloaked against male eyes and the desert climate, two Afghan women walk through a village street. Long, hooded robes have practical value as well as being of social and religious significance in Islamic desert communities. Layers of loose fabric provide the wearer with effective insulation from the desert heat as well as protecting against sunburn.*

Right *A Tibetan woman, on a pilgrimage in the China/Tibet border region, wears a combination of traditional and mass-produced clothing suited to a cold desert climate. The two shirts and the plastic shoulder bag are shop-bought, while the headgear, skirt and quilted jacket are probably home-made. The braided ears of corn are a symbol of the pilgrimage.*

Far right *A group of nomads in Burkina Faso, West Africa. The headdress is an essential component of male clothing in most northern-hemisphere deserts. Serving often as a composite of Sun hat and dust filter, the colour and style of a headdress is often an important indicator of social status. Only one of the group, whose whole style of dress is different, is bareheaded.*

FOOD AND NUTRITION

The range and volume of foods that is naturally available to desert peoples is severely restricted by the availability of water. Because of the intrinsic environmental uncertainty of desert regions, nutrition in deserts and their margins, for both human and animal populations, has always been unreliable and often poor.

In the Middle Eastern and North African desert regions, only livestock rearing and some oasis agriculture are possible. These activities yield a limited range of products: meat, milk, blood and their derivatives; dates and some vegetables, but no staple grains. Diets therefore rich in animal proteins and fats – sheep, goat and camel meat, as well as poultry and eggs, are common, although products from cattle are usually scarce. Pork products are generally forbidden by religion in these regions. Dates are rich in sugar, but diets generally are poor in the starchy carbohydrates and fibre that are derived from bread because grain is difficult to cultivate.

At the desert margins, conditions are less extreme than in the desert cores, and diets are consequently less austere. However, diets are also less stable here because of the great variations in rainfall that occur from year to year. Livestock numbers rise and fall in response to the seasonal and yearly changes in the availability of fodder, which is in turn a reflection of changeable rainfall patterns. Variations in rainfall also affect the "catch" crops of grain or vegetables, which are raised when rains permit. The earliest domestic grain production, probably of barley and later of wheat, can be traced to the desert margins of the Middle East and North Africa 11,000 years ago.

Lifestyle and food

The lifestyles and economic strategies forced on desert peoples are, in general, centred around the continual necessity to insure against food shortage and to preserve what foodstuffs are available. However, in some of the world's desert countries, oil revenues have transformed consumption options as international trade brings in products from all over the world. Since the 1930s, desert areas in industrialized countries, such as those in the Southwest of the United States, have been transformed through the use of irrigation systems into some of the world's major food exporting economies. The people of these areas enjoy a diet as varied and as privileged as any in the world. Similarly, Saudi Arabia has also become a significant desert food producer.

In the absence of such opportunities, nomadism has been the major economic strategy for millennia. Food shortages often follow poor rains, and a mobile way of life allows the desert communities and their economies to be responsive to the ever changing quality of feed supplies. The inherent mobility of this lifestyle means that bulky, heavy food and water supplies cannot be moved about.

The preservation of food is a preoccupation of desert peoples because it is so scarce. Fat and sugar are the main preservatives, and for the most part they are readily available in the desert economy. In the dry desert air, fruit, meat and fish are all easy to preserve, making proteins, carbohydrates and sugars available over long periods. Salt is another important preservative, and occurs naturally in many locations as a result of evaporation.

Left *A Somali woman husking maize (corn) kernels. Brought back from the New World by the Spanish before the end of the 1400s, maize was introduced into Africa by the Portuguese in the early 1500s. Within a generation, the new crop had spread throughout Africa, in many cases supplanting Old World cereals such as barley and millet.*

Far left *Bedouin in Saudi Arabia sitting down to a meal of camel's milk. Although the drinking of fresh camel's milk is widespread among Bedouin, it is more often served in a soured semi-solid form, which is produced by the action of naturally present bacteria. This non-destructive exploitation of domesticated animals (rather than slaughtering for meat) is typical of the desert nomad lifestyle.*

67

TOOLS AND IMPLEMENTS

Inhabitants of deserts live in some of the very harshest environments in the world. Those of the Middle Eastern deserts have to cope with searing daytime temperatures followed by freezing nights. Conditions in the desert regions of northern Asia vary dramatically over the course of one year. In the Gobi Desert, winter temperatures can be as low as −40°C (−40°F) and summer temperatures can soar to as high as 45°C (113°F). In addition, the quality of desert soil is usually very poor and sometimes, as is the case in parts of the Gobi Desert, there is no soil. The lack of soil, even if water were made available, makes it impossible to grow crops or graze livestock. The major problem that is common to all peoples of the desert is, of course, the scarcity of water.

Travelling light

The sparseness of vegetation and the markedness of seasonal changes is reflected by the fact that hunter-gathering and agricultural and pastoral nomadism are usually the only ways of surviving. For this reason, all possessions must be easily portable. This not only disciplines desert peoples against accumulating possessions in the first place, it also ensures that those possessions which they do have are light or disposable. The Australian Aborigines of arid areas, perhaps out of all

desert peoples, travelled the lightest. They usually moved only in small family groups and required little shelter. To protect themselves from the strong winds that sweep the Australian deserts, Aborigines made windbreaks from saplings covered with brush or bark. If it became suddenly cold, they would sleep close to their domesticated dingoes for warmth. When travelling, the men carried simple weapons for hunting, such as spears and boomerangs, while the women carried deep wooden bowls in which to collect fruits and berries. In some areas, plaited or painted bark baskets were carried for gathering, and kangaroo-skin water bags were used in very arid areas.

For the San (or Bushmen) of the Kalahari, a long stick sharpened at one end was used to dig up the roots and tubers that, together with berries and fruits, made up a large part of their diet. While women were responsible for gathering berries and roots, hunting was the preserve of the men. A light bow with poison-tipped arrows was used to kill antelope and small desert mammals.

The nomadic tribes of the Middle East and Asia, compared to the Australian Aborigines or the San, travel in larger groups and need greater protection from the elements. The sparse, or non-existent, vegetation of the Middle East and

Left *An African craftsman at work with traditional hand-tools. Note the spin-drill with a stone flywheel in the foreground of the picture. The metal domes and flower shapes are specialized tools that are used for shaping wooden bowls of differing profiles. The final touches are applied with a conventional tool, while the bowl is held in place by the craftsman's feet.*

Below *A tea-room at the edge of the desert in Baluchistan, Pakistan. Despite their traditional dress, the customers take full advantage of modern implements, such as the wristwatch and factory-produced kettle. The rifle has a triple role in the local culture: for hunting, self-defence and as a symbol of adulthood and membership of a warrior elite.*

Asian desert regions, however, provides few raw materials for people's requirements. In most of these deserts, for example, wood is scarce except for the trunks of palm trees, and even these have few of the qualities of temperate or tropical timbers. Grass and reeds are available in abundance in the river valleys, but these are accessible to only a limited number of these desert dwellers. A few drought-resistant grass species grow in the desert margins and these are available to a more substantial proportion of the desert population. These materials are sometimes used to make objects in which to store and carry food. Although some drought-resistant shrubs are used to make shelters, it is more common for camels, goats, yaks and sheep to provide a source of fibres and skins with which desert dwellers construct their tents.

Changing ways

Modern economic systems, and the engineering, building and transport systems associated with them, have also had a very strong effect on the lives of the peoples living in the deserts of Africa, Asia and the Middle East. The proportions of the populations deriving livelihoods from the desert tracts by nomadic practices have fallen dramatically since the 1950s. At that time, many countries of the Middle East and North Africa, such as Saudi Arabia, the Emirates of the Persian Gulf and Libya, had over 20 per cent of their populations living as nomads. By 1990, the proportions were less than three per cent and the numbers were constantly falling. The places where the craft and creative traditions have survived are those regions where oil revenues have not had a determining effect on local economies. It is in Yemen, in Iran, to some extent in Turkey, in Afghanistan and in the deserts of Central Asia that the traditional ways of life and crafts of the desert peoples can still be found to a significant degree.

Right *In the Danakil Depression of Ethiopia, East Africa, a woman weaves strips of matting from dried grass. The strips will then be sewn together into sheets to make a covering for a framework shelter. When the covering deteriorates, it will be relegated to use as a floormat like the frayed example in the foreground.*

CUSTOMS AND RELIGION

Many people think of Islam as the religion predominantly associated with deserts. Islam is discussed overleaf, but there are many desert regions where the religion of the people has little or no connection with Islam.

In South America, old traditions often mix with new beliefs, and so, for example, Catholicism with an influence from pre-Columbian Inca beliefs characterizes the religion of Peru today. In North America, before the arrival of the Europeans, shamanistic belief systems dominated the desert regions, as they did most other parts of the continent. After the arrival of Europeans, Catholicism again came to dominate the region, although some traditions remained and became absorbed into the new religion.

In Australia, the Aborigines have a rich and varied set of totemic beliefs. In such religions there is a close relationship between a person or group and some natural object or phenomenon, and a kinship between people and nature.

Asia and North Africa

In arid Asian regions, such as Mongolia, Ladakh or Mustang, Buddhism established itself at an early date. It takes the Tibetan Lamaist form. In the Middle East, in outlying areas of Syria, Kurdistan and the Caucasus, the Nusayris, the Yezidis and the Ali-Ilahis have retained customs and beliefs that originated in various cults that characterized the settled Middle Eastern peoples in ancient times. Many of these cults were associated with agriculture, irrigation and a belief in a Mother Goddess. However, the desert, too, housed important centres where, it was believed, the gods had their temples. Two of these places were caravan centres: Petra in Jordan and Palmyra in Syria.

Similar gods and goddesses, and similar temples, characterized the beliefs of the early Arabs in the Yemen, in southern Arabia. Zoroastrianism was the national religion of the ancient Persians and it prevailed in the region until most of its followers converted to Islam in the 600s. There are, however, still many Zoroastrians in Iran. The founder of the religion, Zoroaster (born 588 BC), who conceived of a cosmic struggle between the powers of Good and Evil, was born in the desert region of Chorasmia, bordering on Khorasan, Afghanistan and Turkmenia.

Judaism was a desert religion when the Israelites wandered in the wilderness. Around the date of the birth of Christ, the Jewish sect of the Essenes, and like-minded men such as John the Baptist and "eaters of locusts and wild honey", settled in the Wilderness of Judea near the Dead Sea. Their settlement of Qumran is now well known because of the discovery of the Dead Sea Scrolls nearby.

Christianity in Egypt and Nubia became famous for its desert monks and ascetics and for the establishment of monasteries of the Coptic church. The monks in Wadi Natrun are there to this day, whereas the monastery of St Catherine at the foot of Mount Sinai is all but abandoned. Several desert monasteries were founded in the Judean wilderness outside Jerusalem. At Kilwa, now in the remote Tubayq region of Saudi Arabia, there are to be seen remains of a tiny 6th century settlement of monks.

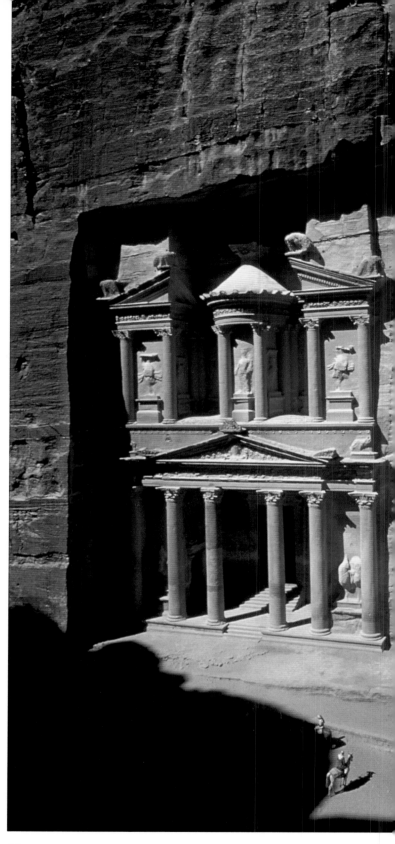

Above A rock-cut facade at Petra, the "rose red city, half as old as time itself". The inhabitants of Petra had their own religion, but were influenced by classical Greek architecture and aesthetics. The elaborate features of the facade, including the decorated pillars, were all patiently carved from the natural cliff face. Inside the cliff are several small chambers. Approached through a narrow, winding defile with sheer walls, Petra was an ideal location for a desert stronghold. A mastery of gravity-fed hydraulic engineering enabled the inhabitants to collect rainwater for irrigation.

Apart from the Arabs, it is the Turkic peoples who have dominated the desert regions of Central Asia. Archeology has shown how mixed were the earlier beliefs of the people of the desert steppes of Central Asia. Excavations at the site of medieval Taraz (Dzhambul) reveal that in the medieval period the surrounding towns professed Zoroastrianism, Christianity and Buddhism. The region had local cults: a Bacchic cult, a cult of the fertility goddess Anahit, and a Turkic cult of heaven. There were also believers in the Manichaen religion, an eclectic offshoot of Zoroastrianism.

Nevertheless, it was the Turkish shaman, a "medicine-man" who assisted the people to maintain the delicate balance between the world of pragmatic necessity and the unseen world of the spirits, who was at the heart of many of the customs and beliefs of the ancient Turks. The religions that later came to dominate the desert regions of Central Asia were deeply influenced by shamanistic beliefs. This was true of Islam; the beliefs became a marked part of the rituals of the mystical movements and orders, such as the Sufis. Many of these orders originated in the desert areas of Central Asia.

Below *A village elder watches over a sacred tortoise in Mali, West Africa. The Dogon people of Mali have a religion which categorizes and ranks all living things. The desert tortoise, which can endure the harshest conditions, is believed to have especial merit, and the tortoise has become a sacred symbol to the Dogon people.*

ISLAM

Above *A traditional mud-brick mosque in Timbuktu, Mali, North Africa, a long-established trading centre in the southern Sahara. Having spread along the Mediterranean coast of North Africa, Islam was carried across the desert by the camel caravans of merchants venturing south to trade fabrics and other manufactured goods for salt and gold with the peoples of sub-Saharan Africa.*

Islam is the dominant religion of the desert stretches of the Middle East and North Africa. Of the pre-Islamic customs that have continued in these areas, some have been absorbed into Islamic practice. Examples include the cult of saints and the belief in superhuman forces, variously termed *jinn*, *ifrit* or *ghul*. These beings are said to inhabit deserted ruins, mountainous rocky outcrops, or solitary trees that have miraculously survived in the wilderness. To the ancient Arabs these trees were sacred. The "burning bush" which Moses, himself recognized by Islam as a prophet, encountered in the Sinai Desert may have been one such tree.

A nomadic religion?

The close relationship between many pre-Islamic customs and Islam itself has prompted the assertion that Islam was a religion born in the desert, a religion that reflected a nomadic mentality. Some historians, however, deny this. They argue that the Prophet Muhammad, the founder of Islam, was a citizen of Mecca, a thriving city supporting a rich merchant class that had little in common with nomadic peoples, and that therefore the desert and its people had little, if any, creative part in the origin of Islam. However, it is also important to remember that it was the same merchant class

that opposed Muhammad, eventually forcing him and his followers to leave Mecca and travel to Medina, where he set up his new religious community. Mecca, however, remains the religious centre for all Muslims. For those Muslims who can afford it, a pilgrimage to Mecca is something that should be undertaken at least once in a lifetime.

Whether or not Islam was born directly out of the desert, the well, the oasis and the mirage furnish much of the imagery of the oral and written literature of Muslim peoples, from Xinkiang to the Atlantic shore.

In the desert, religion and culture are shaped by this environment. Towns are few, and where congregational mosques exist, for example in Timbuctoo in Mali or Bukhara in Uzbekistan, they are substantial structures made of stone or from mud-brick. In open desert, many mosques are simple lines of stones indicating the direction of Mecca. In Central Asia, in the area of former nomadism, there are holy places that are related to the Islamic mystic (*Sufi*) brotherhoods, or to the cult of ancestors (in Turkmenistan), or to pre-Islamic sanctuaries (in Kirghizia) which have since assumed an Islamic mantle. Sometimes they are situated amid Muslim cemeteries. These are not only places for prayer but also, in certain seasons, sites for nomadic fairs and markets.

Islam versus custom

The life of the nomad, whether an Arab, Tuareg, Fulani or Somali, follows a strict pattern based around birth, circumcision, marriage and death, interspersed with seasonal feasts, which are often celebrated with a hospitality shown to both relative and guest. In the desert, women are often freer in many respects than they are in villages or cities, and there are strong matrilineal features among the Tuareg, the Moors, and some of the Tatars. Among the Tuareg, even lineage stems from maternal ancestors.

Often in desert Islamic communities there is to be found a class who form "clerical tribes". The members of this class live in remote areas, teaching the Qur'an in tented schools, curing the sick with amulets, or prescribing remedies that are derived from desert herbs. Their supreme duty is to pray for rain, because only in the coming of rainclouds is to be found the survival of the nomads and their herds of camels, sheep and goats, and – in Sahelian regions – cattle.

Nomadic custom centres around loyalty to the patrilineal (or matrilineal) tribe. It transcends even loyalty to Islam. The great heroes of the desert Arabs, such as Abu Zayd of the Bani Hilal, were prepared to kill and die for their tribe's survival. Abu Zayd crossed the deserts of North Africa to find pasturage for his famished people in Arabia. Unseen voices guided the hero to his destination. The code of hospitality that makes a nomad's life of less value than that of his guest has a potentially negative counterpart that takes the form of a requirement for revenge if the honour of the tribe is at stake. Another area of potential conflict between religious and social duty is that whereas Islam has tried to establish a community of the faithful, custom requires the absolute loyalty of tribal members to fight for the survival of the tribe, its herds and the retention of its pasturage.

Left *A caravan porter kneels in the open desert of the southern Sahara to perform ritual ablutions before prayer. Not even desert travellers are exempt from the requirement of daily prayers and accompanying ablutions. However, Islam is a supremely practical religion, and clean sand may be used instead of precious water when in the desert.*

Above *Discussion outside the doors of the Qaraouiyyin Mosque in Fez, Morocco. Islam is a faith that requires education of its adherents. As a result, desert cities like Fez maintained universities while Europe was still struggling to emerge from the Dark Ages.*

Right *Islam is the world's fastest growing religion with some 400 million adherents. From the early 600s to 750, Islam spread rapidly from the Arabian Peninsula into Persia, North Africa and Spain. Islam then suffered a series of setbacks, but by the 1200s Islam had spread into India, and over the next two centuries large areas of Africa and Southeast Asia were to join the Muslim world.*

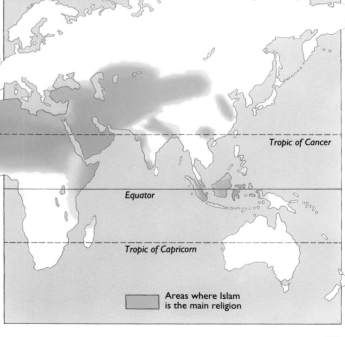

Tropic of Cancer

Equator

Tropic of Capricorn

Areas where Islam
is the main religion

TOWNS AND CITIES

Left *Traditional houses in the Asir Province of Saudi Arabia. Although these mud-brick buildings look elaborate and give an illusion of permanence, they require constant maintenance if they are to remain habitable. Despite the taller houses (with glazed windows) showing obvious signs of occupancy, some of those in front appear to have become semi-derelict.*

People were slow to build settlements in deserts. The earliest communities faced the major problem of finding reliable water supplies for drinking and cultivation. Sometimes, because of their extreme isolation and difficult access, desert settlements were built as refuges for people who had been driven from more fertile lands by invaders.

In the Nile Valley, Mesopotamia, the Iranian Plateau and the Lower Indus Valley, settlement began 6,000 years ago. The Egyptian and Mesopotamian settlements depended on efficient irrigation systems which, along with the fertile soil, brought in huge harvests. Soon, in areas such as those around Edfu and Luxor on the banks of the Nile, there evolved a sophisticated urban culture. Other cities flourished in desert conditions in and around the Nile Delta. The fabulous cities of Ur and Babylon grew up in Mesopotamia, and Mohenjo-Daro and Harappa arose in what is now Pakistan. Urban life has continued in Egypt ever since, despite occasional political and economic upheavals. Settlements in Mesopotamia, however, foundered through soil salinization although what are now Baghdad and Damascus remained important centres of population throughout most of the historic period.

The early Persians used river water and underground canal (qanat) systems to support large desert towns. The empires of Cyrus and Darius, for example, evolved in the arid plateaus of the Iranian Desert at sites such as Parsagard and Persepolis. Although between 336 and 330 BC these centres were destroyed by Greek and Arab invaders, Alexander the Great set up governorates (*satrapies*) in outlying areas within the Persian Empire which maintained an Irano-Greek urban form for many centuries after his death. Balkh in the Turkestan Desert of Central Asia is an example of one such governorate. The Mogul emperors in India, similarly, took forms of Persian urban culture to desert areas of India, notably the Thar Desert of the northwest.

The influence of Islam

With the rise of Islam in the AD 600s, Mecca and Medina became important centres of pilgrimage in the Arabian Desert, which, until that time, had lacked significant settlements. Islam not only revitalized many towns and cities in both the Nile and Tigris-Euphrates valleys and elsewhere in the desert lands of North Africa, the Middle East and South Asia; it also played an important part in the development of new towns such as Cairo, Marrakesh and Fez, with their elegant minarets, mosques, universities, palaces, bazaars and city squares. Today, North African, Middle Eastern and South Asian areas remain some of the most densely populated desert regions anywhere in the world.

Before the Arabs, very few cities grew up in the deep Sahara. Small settlements such as Ghadames and Murzuq were all that could survive on limited water supplies in so hostile an environment. The semi-arid Mediterranean fringes of the Sahara, however, supported magnificent early cities.

The invasion by the Arabs of North Africa in the early centuries of Islam created a new series of towns in the mountain rims of the Sahara in North Africa. They were populated by Berber-speaking tribes driven out from the Mediterranean coastal strip by the Arabs. The Berbers created hilltop fortresses as a defence against attacks by the new occupiers of the region. In places as far distant as Gharyan in Libya and Matmata in Tunisia, Berbers adapted to desert life in troglodite settlements, in which people built their houses below ground. By the 1350s, trans-desert trading stations grew up in the south of the Sahara; among them Timbuktu became one of the most important.

New World settlements

The Spanish conquest of the Americas in the 1500s resulted in the establishment of many colonial ports and administrative centres, many in deserts such as the cold Patagonian Desert of South America and the hot deserts of California. Urban forms were taken from Spain and were dominated by military and religious influences imported by the Spanish colonial authorities, particularly from Andalusia. In later centuries, the towns in the American deserts took on regional characteristics of their own, particularly those in North America, although Spanish influences are still apparent through the development of sites chosen by the Spanish and in the names of towns such as Los Angeles and San Francisco.

Left *Granaries built on a steep hillside by the Dogon people of Mali, West Africa. Too small and crowded to be dwellings, these permanent granaries are built from stones found on-site, bound together with mud and clay. Some of the granaries have the typical flat roof of desert buildings, while others have a conical thatch associated with wetter climates.*

Above *Cliff dwellings at Mesa Verde National Park, Colorado, USA. These houses were built of rock and adobe (Sun-dried bricks) by the Anasazi people, between 800–1000 years ago. The cliff overhang provides an easily defensible location that is protected from the elements. Today, the Hopi people, descendants of the Anasazi, live in similar communities.*

GROWTH WITH WEALTH

Large cities have existed on many desert margins for thousands of years. Many, such as Damascus or Mecca, were important trading or religious centres. However, large-scale urbanization, particularly of Arabian and North American deserts, only occurred when mineral exploration revealed the wealth of many desert regions, and modern water management techniques made it possible to extract and transport sufficient quantities of water to satisfy the needs of large modern cities. Migration from rural areas to towns, particularly in developing countries, has also played an important role in the explosive growth of cities.

Liquid gold

Oil, more than any other commodity, brought rapid and sustained urban growth in the deserts, first in North America and later in Arabia and the Gulf in the Middle East and Saharan North Africa. Oil exploitation required new mining and service towns, such as Abadan in Iran and Dammam in Saudi Arabia, which were located in the oil fields themselves. Oil brought great wealth to formerly poor desert economies, and an era of rapid and extensive town building began throughout the Arabian, Iranian and North African deserts in the 1960s, exploding during the 1970s in response to the massive increase in oil revenues. Large-scale construction occurred not just in the oil-producing states themselves. Countries such as Jordan and Egypt, which supplied oil transit facilities, labour and political support, also underwent rapid urban expansion.

Provision of services for desert settlements is expensive and in the oil-producing areas can be supported only because large revenues are earned from oil exports. The population of Kuwait City, for example, is reliant for as much as 90 per cent of its income on oil. A failure, or gradual fall, in oil revenues would make many of the urban settlements of the oil-exporting world entirely unsustainable.

Above A mosque on the Corniche Road, a prestige street in Abu Dhabi, United Arab Emirates. Abu Dhabi is typical of many of the new desert cities built by oil wealth. Modern concrete and glass buildings abound, even when shaped into traditional designs. In many cases, old and historic buildings have been demolished to make room for modern constructions.

Left A street in the souk (market) area of Marrakesh, situated on the edge of the Moroccan desert. Morocco lacks the oil wealth possessed by many other Arab states and, as a result, its cities still have a traditional appearance. Strip away the lights, cables, and advertisements of the present century, and this street scene has a timeless quality.

Tourist towns

Tourism of all sorts has flourished in many desert regions outside the United States. In the heart of the Sahara, for example, and at Aswan on the Nile River, high-grade hotels provide winter sun for tourists from Europe, as well as access to the sites of Ancient Egypt.

At Agadir, on the western rim of the Sahara in Morocco, the hot, sunny climate is exploited in a purpose-built tourist town. Tunisia has also benefited from the growth of the tourist industry, and its resort towns now welcome thousands of sun-seeking northern Europeans every year.

At Uluru (also known as Ayers Rock) in central Australia, tourist complexes now accommodate increasing numbers of visitors to this spectacular sacred site.

A developed economy

In the complex economy of the United States, settlements in deserts have been used for a great variety of purposes. Much of the North American desert and the surrounding semi-desert was not attractive for farming during the westward expansion of the United States. By the 1950s, however, dams controlled the flows of major rivers and modern irrigation supported rich orchard and arable estates, and made possible the growth of cities such as Las Vegas and Phoenix.

Rising standards of living after World War II brought a life in the Sun Belt – in states such as Colorado, California, Texas, New Mexico, Arizona and Nevada – closer to the pockets of many more people. The trend was led by older people seeking warmer places for their retirement, where sports could be pursued in reliable weather conditions. Tourism and recreation became an increasingly important industry in the Sun Belt. Industrial towns grew up because the dry climate suits certain industries, and the sunshine is popular with the workforce. The rapid growth of Phoenix, Arizona, reflects these developments. Phoenix's water supply was originally ensured by the building of the Theodore

Roosevelt Dam in 1911. But by the 1950s, mainly because of the introduction of air conditioning, the population rapidly grew, and the city's area increased from 44 square kilometres (17 square miles) in 1950 to more than 971 square kilometres (375 square miles) by the late 1980s. This expansion made necessary the construction of the Central Arizona Project, which diverts water from the Colorado River and began supplying Phoenix in 1986.

This kind of growth has placed unsustainable pressures on the natural environment. In the Central Valley of California, and in many other places, shortages of water for urban use are reflected in periodic rationing of water supplies. The cost of providing water in desert areas is rising worldwide. There is competition for water between towns and the countryside, in which water is increasingly being bought by towns, and so the share of water available for farming decreases. Meanwhile, pumping of underground aquifers to supply urban needs has depleted fossil aquifers almost to extinction. Around the Great Sand Desert of Iran, for example, many villages, having pumped out their sub-surface supplies, must now survive on low volumes of brackish water.

DESERT RESOURCES

Regions of many deserts are rich enough in resources to sustain the lives of large communities, although in the past this has only been in unusually favoured locations, such as major river valleys. Until recent times, the most valuable resource for desert peoples was water. New technologies, however, have made it possible to exploit mineral deposits hidden beneath many of the world's deserts. The most obvious example is that of oil in the Arabian Peninsula, a resource so great that it has turned some of the world's poorest countries into some of the richest. Extraction of resources can cause untold damage. Opencast mining removes great areas of land; all forms of mining create large amounts of waste; and the processing of ores can produce deadly toxins.

Above *Flaring gas at Hassi-Messaoud, Algeria.*
Left *A tall drilling rig gleams under desert skies in Texas, USA.*

THE RICHES OF THE DESERT

Throughout history, deserts have been regarded as empty, inhospitable places. To all but an enterprising few, they have been barriers to movement and trade. People have had to develop survival strategies based on the conservation of very scarce resources. In recent decades, however, the extraction of hydrocarbons and other minerals has brought about the transformation of some desert societies.

The traditionally used desert resources are vegetation and water. Most of these are renewable, including major rivers, some groundwater resources and pasture. Some are non-renewable, notably sub-surface groundwater resources.

Deserts have been minor parts of the world's agricultural economy, but new technologies for water management have allowed some desert areas, such as southern California, to produce abundant fruit and vegetables. However, for most desert peoples, livestock rearing, or pastoralism, has long been the only viable economy. In the late 1900s, the need to feed rising populations has played a part in leading many desert states to adopt unsustainable agricultural policies. As a result, water and soil have been both diminished and degraded. In some regions, modern extraction and irrigation technologies have severely depleted groundwater, or led to the salinization of the soil.

Non-renewable water resources have become increasingly important in desert regions. Most of these are probably water accumulated in the past 100,000 years. Since the 1930s, the search for crude oil in the Middle East, Central Asia and North Africa has led to the discovery of much of this water and its subsequent exploitation.

Above *A Tuareg farmer poses in front of a newly installed solar-powered pump near Timbuktu in Mali, West Africa. Modern desert farming is often a combination of old and new techniques. Although the water is brought to the surface by a high-technology pump, the means of distributing water to the crops appears to be traditional.*

Right *Part of the world's largest solar-power plant at Luz in the Mojave Desert, California, USA. Desert regions have enormous potential to benefit from abundant sunlight. The development of a cheap method of mass-producing electricity directly from sunlight could transform the economies of many desert areas that lack substantial oil and gas deposits.*

Far right *A human chain of Boran people lift water from the Singing Wells at Marsabit in northern Kenya, East Africa. The water is poured manually into an irrigation channel that carries it to the village fields. Even with existing technology, a small solar pump could perform a task that currently requires the labour of nine adults.*

Energy and minerals

In economic terms, the most important resources of desert regions are crude oil and natural gas. Past climates created forest and swamp environments in which residues of decomposed vegetation were laid down, eventually forming crude oil and natural gas. Some states with large crude oil deposits have become fabulously wealthy: Kuwait, Saudi Arabia and the United Arab Emirates enjoy the highest gross national products per head of any countries in the world. Oil has also indirectly benefited both regional and global economies, giving employment to many millions of workers in oil-rich countries. At times, around 20 per cent of Egypt's gross domestic product has been generated by remittances sent home by citizens working in the oil states.

Industrial activity is not extensive in desert economies. Until the 1980s, most crude oil was exported rather than processed near the source. The first oil crisis, in 1973, enabled oil-rich governments to invest in plant and equipment, and during the 1980s desert economies won an increasing share of oil-based energy and chemical production.

Deserts are also important repositories of non-renewable minerals. The Sahara is one example, containing important iron ore and phosphate deposits. The deserts of Iran are rich in minerals, including copper, those of North America and South America in precious metals and those of Australia in a wide range of mineral ores.

The major renewable resource of desert regions is solar energy. To date, the only large-scale use of solar energy has been to heat water in homes. These systems have been widely adopted in Israel and Jordan, as well as in the conservation-conscious states of the southern United States. Tapping the full potential of the sun's energy requires more advanced technologies to convert the energy into useable forms.

Desert economies are deficit economies

The renewable natural resources of the deserts are subject to degradation, while non-renewable ones are diminished by use. Neither is equal to the challenge of sustaining the current population. Consequently, food and other commodities must be imported. Although the extraction and export of crude oil and its derivatives are beneficial in the short term, they are based on the progressive depletion of national assets. Each year, desert economies are depleting natural resources, without establishing balancing value-adding activity to secure the livelihoods of succeeding generations. The move to industrialization which took place in the 1980s is one of the few strategies which might secure the future. Others include the investment of oil revenues on international capital markets, and involvement in financial and banking services.

HERDING AND RANCHING

Livestock rearing is the main livelihood for many of those desert populations that do not have access to water sufficiently abundant to sustain agriculture. Nomadic livestock rearing is essentially a low-intensity activity, and as a resource-using strategy can be consistent with the limited forage naturally available in desert regions. People and animals need constant supplies of water and food; in desert areas there is by definition very little water, and unreliable rains can reduce vegetation cover to almost nothing. In order to make best use of the natural forage that is available in various places at different times, most desert herders and their families are mobile, responding to seasonal changes in temperature and rainfall. These movements may follow particular changes in the weather, but more often follow regular annual cycles.

Traditional strategies
Livestock rearing is centred around the desert margins, and is carried out mostly in areas that receive some 50–200 millimetres (2–8 inches) of rainfall annually. However, herders have traditionally been able to gain occasional access to grazing in regions which receive on average more than 200 millimetres (8 inches) of rainfall per year.

The margins of the desert are prone to drought. During major droughts, the herders reduce livestock numbers and, in the past, used the spare grazing capacity of better watered marginal areas. In recent years, population growth and the intensification of soil and water use for permanent agriculture throughout the settled areas have severely affected the ability of such tracts to absorb the occasional migration of drought-afflicted herders and their livestock. As a result, livestock rearers have had to seek supplies of supplementary feed. This is both to meet the annual seasonal deficit that occurs during the dry season and to relieve the accumulating stress of the shortage of feed that is caused by increasing human population and a growth in livestock numbers.

A feature of traditional livestock rearing is the tendency for herd sizes to rise quickly after a period of stress, and to fall dramatically during periods of drought. However, in those areas where the demand for livestock products is high – because consumers have access to oil revenues or enjoy high urban incomes – the very high prices that can be charged for livestock products can lead to overstocking of the fragile rangelands. In the long term, the pressure on natural vegetation is unsustainable and severe damage can be caused to the ecosystem, reducing its productivity.

Right *Cattle collected behind fences at the edge of the Kalahari Desert in Botswana, southern Africa. Nomadic pastoralism does not convert easily into modern ranching. These cattle, for example, cannot forage for their own food in these desert conditions, and must be fed on scarce fodder that has been harvested elsewhere, requiring additional transportation.*

Below left *Goats around an ancient, stone-walled well in Somalia, East Africa. In the background, other animals are enclosed by a fence made of dried thorn bushes. Although goats are hardy enough to withstand desert conditions, they must be carefully controlled; otherwise their voracious feeding will consume all the available vegetation, adding to the risk of desertification.*

Above *Camels being watered from a shallow well in Niger, West Africa. Camels are the major livestock resource of true deserts, being very well adapted to their harsh environment. Thousands of years ago, the domestication of the Arabian camel (dromedary) enabled humans to cross and populate the* Sahara and Arabian deserts, and brought about trans-desert trade. The Bactrian (two-humped) camel, played a similar role in the human occupation of the cold deserts of Central Asia and China. Today, the camel is completely domesticated and has not existed in a wild state for hundreds of years.

Domesticated animals

The camel is the major livestock resource of the deserts of northern Africa and Asia. It is a hardy animal, able to withstand heat and the absence of water for long periods. In addition to providing livestock products, the camel has also traditionally been the major transport animal. For centuries, the camel made trans-desert trade possible, even across such vast, inhospitable regions as the Sahara. Like the camel, the goat is a robust animal well able to withstand the extreme environment and poor forage of the central desert tracts. Sheep are the major economic livestock in dry regions, and the meat and other products which they provide are highly prized. Cattle are generally not suited to desert conditions, although they are a feature of the agriculture of river valleys in desert regions. Here they are used for draught purposes, for transport, and for milk and meat.

Because of the increasing populations of the world's deserts and their margins, the demand for livestock products has increased steadily over the past century. In North Africa, further increases have been exerted by the growing and increasingly prosperous settled populations of the Nile valley and the Mediterranean margins. This has led to larger herds, maintained through the provision of wells and supplementary feed. This has led to severe overuse of natural pastures.

Modern farming has deployed high technology to support livestock rearing in desert areas. Irrigated grain and alfalfa production in the Kufrah region of southeast Libya, and intensive irrigation schemes around Riyadh, have been used to support large populations of cattle and sheep. In terms of the depletion of natural resources, the costs of such projects have been immense. As a result, the experiment in southern Libya was abandoned by the end of the 1980s, although the Saudi Arabian schemes have been maintained.

TRADITIONAL FARMING

The earliest civilizations were founded in desert areas where rivers and springs provided water and a basis for an agricultural economy. On the major rivers that traversed the deserts of northeast Africa, the Middle East and northwest India – the Nile, the Tigris-Euphrates and the Indus – the combination of a reliable supply of water and good growing conditions stimulated farming traditions which have, in places, endured for 6,000 years or more. Irrigation enabled unprecedented levels of crop production.

The traditional irrigation of the Nile Valley made use of the annual flood which reached Egypt in the late summer. The Nile waters also brought silt from Ethiopia which annually enriched the fields of the narrow Nile floodplain and the delta. After the waters fell away, water had to be raised during the later part of the season. This was achieved by a variety of ingenious lifting devices, including the *shaduf*, a pivoting "see-saw" device, and the Archimedes screw.

Elsewhere in the arid world, particularly in Persia, ancient technology also secured the basis of reliable agricultural production. The Persian wheel lifted groundwater from depths of up to 20 metres (66 feet), and was then used to lift the water from channels up to the level of fields. The system was adopted over many of the arid and semi-arid regions, from northeast Africa to northern India. The precise origin of the other invention of great hydraulic significance associated with Persia, the qanat (also known as the *falaj* in Oman, and the *foggara* in North Africa) is not known. It has been most extensively used in Persia (modern Iran), and examples of the technology have been observed in Algeria, Tunisia, Pakistan, China, Libya, Syria, Afghanistan and even in Japan. These laboriously constructed, graded underground tunnels conveyed water sometimes tens of kilometres from distant groundwater resources to dry tracts that enjoyed good soil and other favourable conditions for agriculture.

Above *Waterwheels used for lifting water in Egypt. In this traditional irrigation method, the wheels are turned by the flow of water from a main irrigation channel diverted from the Nile River. Cups attached to the sides of the wheel lift water from the main channel and tip it into a higher-level channel for distribution to fields. Wheels such as these have been in use since Roman times or earlier.*

Respecting resources

Most of the ancient technologies, when compared to modern methods, were intrinsically sustainable in terms of water use. The operation of traditional water-lifting devices required the deployment of a significant amount of energy, which could only be supplied by either animal or human power. The rapid depletion of the water resource was automatically restrained by the costs of raising and moving water. Even the systems that used gravity to transport water were built in accord with principles of sustainability in that the builders well understood that there was little point in constructing a qanat, at massive cost, that would use water beyond the long-term capacity of the distant aquifer.

The prosperous towns and villages of arid Persia that were once supplied by the sustainable subsurface qanats have been seriously affected by the introduction of modern pumping technology. This can raise much larger volumes of water than supplied by the traditional systems, and in so doing can deplete groundwater at a much more rapid rate, making the ancient hydraulic systems redundant.

The rehabilitation of the ancient systems becomes impossible when either the physical or social infrastructure upon which they depended has been damaged or altered.

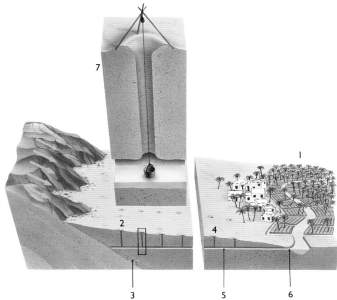

Left *The ruins of an abandoned town at Timimoun Oasis in Algeria, North Africa. Because of changes in the hydrology of the region, the oasis could no longer support the town and its population.*

Below left *A farmer uses his hoe to open a temporary sluice in the mud wall of a field near the banks of the Nile, Egypt. This type of "soft" engineering is typical of water management on flood plains.*

Above *Qanats carry groundwater for irrigated areas (1) without evaporation. When a qanat is built, first a headwell (2) is sunk down to the water table (3). This well may be 100 m (330 ft) deep. A line of ventilation shafts is dug (4), and then an underground channel (5) is begun from the qanat mouth (6). Gravity moves water to the mouth where it is needed, and water can be drawn from the shafts (7).*

When groundwater levels are driven down by excessive pumping, qanats cannot deliver water. But just as important are social changes, such as economically driven emigration which impairs the availability of labour and of finance to maintain the tunnels. Rehabilitation, however, has been possible in some of the Omani qanats and South American systems. The open canals of the Patacancha Valley in Peru have been restored and are providing water.

Rainfed farming

As paradoxical as it may sound, for thousands of years some desert peoples have relied on rain to grow crops in arid regions, particularly where irrigation is not possible. The levels of production are determined by the rains of a particular season. The peoples using these opportunistic techniques do not depend on the crops for their livelihoods. The crops are raised when possible and hence are known widely as "catch-crops". Barley and wheat are sometimes grown in this way in the deserts of North Africa, the Middle East, and Southwest and Central Asia. Rainfall of about 200 millimetres (8 inches) is required to ensure a good crop of grain or straw. If the rain is insufficient, the crop will not mature, but often it can still be used to graze livestock.

MODERN WATER STRATEGIES

The human and animal populations of arid areas have been expanding rapidly since the beginning of the century, especially in the past 40 years. In the past, population levels were naturally regulated by periodic famines, and even today, some of the peoples living on the margins of the world's deserts endure extreme hardship. Natural resources augmented by traditional irrigation technology have not provided food security. The development of water resources by modern techniques of water management has been seen as a way to solve this problem.

For thousands of years, dams have been built in arid regions. The Marib Dam of Yemen and the structures built during the Roman period throughout the Middle East and North Africa are among the finest examples. These structures were often temporary and created few, if any, environmental problems. Since the beginning of the 20th century, however, large structures have been built throughout the dry areas of the world to control the supply of water. At first, these immense structures were built for agricultural purposes, but they were soon developed to provide hydroelectric power. In this way, ever greater volumes of water were diverted from their original courses, often resulting in the destruction of ecologically important floodplains.

Other kinds of mismanagement affects the river valleys of the Old World deserts because the river basins are divided between more than one country; the Nile, the Tigris-Euphrates and the Indus are particularly affected. These rivers are fed by water from upland Ethiopia, the East African Highlands, the mountains of Turkey and Iran and from the Himalayas. Because water loss through evaporation is lower at higher altitudes, dams should be sited in these upland regions to provide more economical storage than if they were built in true deserts. Such dams, however, affect the natural water cycle of the rivers.

An irreplaceable resource
Water stored underground is safe from depletion by evaporation. Groundwater occurs naturally beneath vast tracts of the deserts of the world and such water is often of sufficient quality for human consumption and for irrigation. A high proportion of the groundwater of the Asian and African deserts is "fossil" water that accumulated over the past 100,000 years. These ancient and generally finite groundwater resources have been severely depleted throughout the deserts of the world since the 1960s and 1970s. In the United States, pumping and water distribution technologies were developed and widely deployed to exploit groundwater in the southern states of Texas, Colorado, Arizona and California. This led to massive extractions of groundwater and to a rapid decline in water levels. The result was increased water costs which made a great deal of farming economically as well as ecologically unviable.

Because deserts can bloom if water supplies are sufficient, engineers have, over the years, evolved ways of transporting water from neighbouring water-rich areas to desert regions. Vast new tracts were irrigated by unlined channels during the second half of the 19th century and the first half of the 20th century, enabling the increasing populations of desert countries such as Egypt, Iraq, India (including the present Pakistan) and Central Asia to be employed and fed.

These inter-basin transfers have been extremely important in terms of the volume of economic activity generated. Since the 1950s, water transfers between basins have been increasing in significance. In the United States, transfers of water from the Colorado Basin to southern California are crucial to the desert economy of the recipient state. However, arid northern Mexico has been adversely affected by these water exports from the Colorado Basin because they reduce flow in the Colorado river. In turn, Mexico itself has plans to engineer transfers of water from its well-watered western and southern mountains to the arid north of the country.

The dangers of such huge water transfers have become apparent in the former Soviet Union. Here, the agricultural development of the arid south was to be based on the transfer of water from the rivers of the humid north. Only part of the scheme has been completed; the consequences for the Aral Sea have been disastrous, and the livelihoods of local people and the local ecology have been ruined.

Above The Hoover Dam, located on the Colorado River on the border of Arizona and Nevada, USA, is typical of the water management schemes undertaken in the first half of this century. Although it provides much of the Pacific Southwest with electricity and irrigation water, it has reduced the amount of water reaching consumers in Mexico.

Right Large-scale spray irrigation of desert farmland in Utah, USA. Throughout the world, the use of fossil groundwater for agriculture, particularly during the last 40 years, has transformed arid regions into valuable farmland. Losses because of evaporation are high, however, and this resource is being seriously depleted.

RESHAPING THE LAND

New technology has made it possible to level land more extensively and more quickly than in the past and to cultivate more thoroughly and more often. At the same time, water technology has enabled people to lift and move massive quantities of water over great distances, as well as to distribute it to vast irrigated tracts. The ability to manage soil and water resources in these ways has had many very powerful impacts on the surface and sub-surface of the world's deserts, especially at their margins.

The capacity to disturb the land surface and to remove the natural protective shrub and herb cover has led to the degradation of extensive areas of the world's deserts. The Middle East and North Africa have many tracts of dry land that have been irreversibly damaged by attempts to undertake the cultivation far beyond that which could be sustained by existing levels of rainfall. Extensive parts of Iraq, Syria and Jordan have been seriously degraded, and in Libya, Algeria and northwest Egypt, marginal land has had its agricultural and ecological potential destroyed though soil erosion caused by such attempts at cultivation.

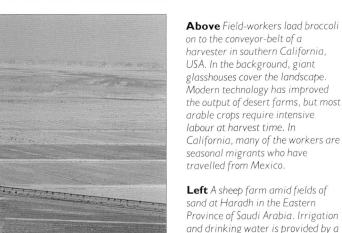

Above *Field-workers load broccoli on to the conveyor-belt of a harvester in southern California, USA. In the background, giant glasshouses cover the landscape. Modern technology has improved the output of desert farms, but most arable crops require intensive labour at harvest time. In California, many of the workers are seasonal migrants who have travelled from Mexico.*

Left *A sheep farm amid fields of sand at Haradh in the Eastern Province of Saudi Arabia. Irrigation and drinking water is provided by a gridwork of steel pipes. Fodder harvested from the fields is delivered by tractor and trailer. Part of the flock has been let out for a rare treat, the opportunity to graze for their food, if only on stubble.*

Top left *Young avocado plants on an agricultural kibbutz in Israel. A network of micro-bore tubing delivers minute quantities of water (trickle-drip irrigation) to each plant. The smaller plants are still surrounded by mesh guards, which prevent the newly emerged leaves from being shrivelled by the strong desert sunlight.*

Greening the deserts

The increased use of water that has been made possible by modern techniques can improve agricultural production, but rarely increases economic productivity when such things as costs and investment are taken into consideration.

The greening of desert areas for both irrigated farming and the amenity of gardens and trees in both the countryside and the city is generally welcomed. The desert countries of the Gulf have invested heavily on improving the appearance of whole cities and of long stretches of highway. Abu Dhabi, for example, has achieved such a remarkable level of "greening" in its major urban centres that they can now almost be regarded as garden cities.

The use of high volumes of water can at the same time have a number of serious negative consquences. All water contains some salts and so can cause salinization of soil, especially when evaporation is high. Such salinization is not a new phenomenon. Ancient irrigation schemes faced similar problems, but the pace of deterioration tends to be quicker now because modern water management techniques have made possible schemes on an unprecedented scale.

Ineffective water management also leads to poorer groundwater quality. Water that passes through a layer of soil tends to accumulate salts, and if there are large amounts of salts in the soil, salts in that water can reach high concentrations. Salinized water that moves down to the water table pollutes the groundwater resource and restricts its further use in agriculture.

Careful management

Despite all these problems, modern technology can also provide very important benefits for the efficient use of the scarce water. Indeed, the successful management of the world's deserts is, to a considerable extent, a matter of the effective management of technology. Metering systems can assist farmers and project managers to regulate the use of water. Such regulation can be automated. In one kind of system, sensors are distributed through an irrigated area and pass information back to a computer that controls irrigation; the computer ensures the optimum availability of water to crops while at the same time minimizing its use. At the level of a small farm, modern water distribution systems known as "trickle" or "drip" systems deliver water in carefully controlled amounts; such systems can increase the efficiency of water use by over 100 per cent for crops that can be grown in rows and for tree crops.

Whether modern technology has positive or negative effects on the surface and the subsurface depends on the manner in which it is used. Technology can be environmentally appropriate when properly managed. When it is used carelessly, its full destructive capacity to disturb and irreversibly damage soil and ground water is brought to bear. Sound water technology must be the basis of the future utilization of the world's deserts. It will, therefore, be very important to develop effective institutions at all levels – from national governments to individual farms – to deploy and regulate the use of these powerful technologies.

REMOTE GROUNDWATER

Modern drilling, pumping and distribution systems have made it possible to lift groundwater in remote desert areas from depths as great as 500 metres (1,600 feet). Once out of the ground, the water can be applied over enormous areas according to schedules that meet the irrigation requirements of a wide range of crops.

The main technology to have developed relatively recently for utilizing groundwater in agriculture is the centre-pivot structure. Water is conveyed under pressure from the central source by means of an electric pump. It then flows along a moving gantry, which circles around the source distributing the water over the crops. The speed at which the equipment circles is regulated as necessary to determine the frequency of application of the water.

The result of the centre-pivot system is a dramatic circle of vegetation in a barren sea of sand. A major advantage of the system is that it can be operated on quite uneven land, which means that the high costs of land levelling can be avoided. On the other hand, it has been found in the United States – where centre-pivot equipment is widely used for supplementary irrigation in the dry southwest and in some of the arid mountain states – that better germination and growing conditions occur when preparatory levelling has been carried out. The main problem is that on irregular surfaces water tends to accumulate in the hollows and drain from higher parts, leading to uneven access to water and hence variations in crop development.

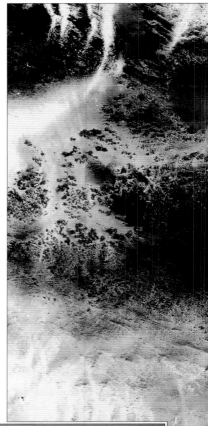

Right A picture generated by an imaging satellite showing irrigated fields near the Kufra Oasis in Libya, North Africa. The distinctive circular shapes produced by centre-pivot irrigation equipment are clearly visible. Variations in the apparent colour of the fields indicate crops at different stages of growth.

Below An aerial view of irrigated fields in the Libyan Sahara. Each field is about 1 km (0.6 mile) in diameter. The pale strip running across the nearest field is a road providing access to the pump and pivot equipment located at the centre. Although a triumph of modern agricultural techniques, such projects are very wasteful of fossil water resources.

Theory into practice

In 1971, Libyan officials were advised that the high-quality groundwater of the Kufrah Basin would enable the production of wheat to help the country become self-sufficient in this staple crop. Unfortunately, the wheat project was not successful, and it was decided by the mid-1970s that a 10,000-hectare (25,000-acre) livestock scheme, based on the production of alfalfa, would be a preferable strategy. Difficulties arose, however, not because of poor availability of water or inadequacies in the soil, but through the harmful effects of hot dry winds on plants and livestock, and ultimately because of the health of the sheep flocks could not be adequately maintained.

Also in Libya, an even larger scheme covering some 50,000 hectares (125,000 acres) was attempted in the Sarir Basin, drawing on another resource of ancient groundwater. Again the project encountered problems, not the least of them social. The organizational and technical difficulties led the Libyan government to switch to a strategy of piping water from the southern aquifers to the coast, which is where the users of the water preferred to live.

The most extensive deployment of the centre-pivot system outside the United States has been accomplished in Saudi Arabia. A wish to achieve food self-sufficiency combined with seemingly unlimited supplies of capital from oil revenues brought about a government-led policy to utilize the groundwater of the Riyadh region. Many hundreds of thousands of hectares (acres) were developed for irrigated farming. By 1990, Saudi Arabia was raising a significant proportion of the country's animal feed requirements as well as becoming, rather surprisingly, an exporter of wheat and a number of other agricultural products.

The sustainability of the desert irrigation projects around Riyadh is, however, very much in doubt. They are not economically secure because the output is only achieved at many times the world price for the commodities produced. Moreover, the groundwater resource, which is only partially renewed each year, is being rapidly depleted.

A free commodity

The centre-pivot systems, although relatively water-efficient compared to other methods of irrigation, are not ideal in terms of water conservation compared with various drip systems. The latter, however, can only be used with a narrow range of tree and row crops. As a result, drip systems are not so popular, especially where the dangerous perception prevails that water is free.

Governments throughout the arid world have difficulty managing a resource that has been traditionally viewed as free, albeit precious in terms of availability. Attitudes are difficult to change, especially once it seems that water has become more abundant. Most government bodies are loath to argue for ecologically and economically sound water resource allocation in the face of elite organizations and professional interests, which argue for the development of water for farming in the misguided belief that it will create long-term wealth and self-sufficiency.

OIL AND GAS

The world's oil industry dates from 1859, when Colonel Edwin Drake began processing oil from shallow seepages in Pennsylvania, in the United States. The industry grew rapidly as American oil companies sought new, easily exploitable reserves. In particular, their search took them to the desert areas of the west and southwest of North America, and major discoveries were made in the arid regions of Texas and California. The desert areas of Mexico were also explored and developed for petroleum at an early date. Deserts were easy for the oil companies to explore and develop. The absence of vegetation made geological surveying straightforward, and there was little existing industry or agriculture to compete with oil exploitation.

The process of exploration began seriously in the early 1900s when the world's largest resources were discovered beneath the deserts of the Middle East and North Africa. In 1908, oil was discovered in Iran at Mesjed-e Sulaiman on the fringes of the Khuzestan Desert by a company that eventually became British Petroleum. Elsewhere in the Persian Gulf region, exploration was spurred on by the finds in Iran. Oil was soon found in Iraq, Saudi Arabia, Kuwait and Bahrain, though development was delayed by the advent of World War II. At the same time, the Sahara, the Australian Desert and the deserts of South America were all explored.

Oil was later discovered in the Libyan Desert as far south as the Calanscio and Murzuq Sand Seas, in Algeria at Hassi Massoud, and in the cold deserts of Alaska and Siberia. But the Iranian, Arabian and Iraqi deserts remained the most concentrated areas of oil and natural gas reserves. By the end

Inset left *A worker adjusts a valve on an oil-pipeline near Ahwaz in southwestern Iran. In the background, two flares burn off natural gas, which is an unwanted product at this particular field. The crude oil is pumped, via the pipes seen in the foreground, directly to loading facilities at tanker terminals on the Persian Gulf about 200 km (125 miles) away.*

Above *An oil rig sited on dunes in Oman. The desert is a difficult place in which to work. Before any drilling can commence, a stable surface must be created amid the shifting sands. Throughout the lifetime of the rig, wind-blown sand is a machinery-clogging nuisance, and creeping dunes threaten to engulf the whole installation unless protective measures are taken.*

of 1991, the Middle East has been estimated to have no less than 66 per cent of all world crude oil and 30 per cent of natural gas reserves. Oil production in this area was 26 per cent of the world total. In North Africa, Egypt, Algeria and Libya accounted for almost 6 per cent of world oil output.

The history of the oil industry in the deserts is remarkably short. But the future of the industry may be equally short. Ratios of reserves to production rates indicate that many oil fields will soon reach the end of their useful lives – about 10 years in the United States, 20 years in Oman, Qatar and Bahrain, and less than 50 years in many other areas. However, reserves in Saudi Arabia, Iran, Iraq and Kuwait are so large that they may last much longer than this.

The short life of the oil industry contrasts sharply with its enormous ecological and human impact. The oil and related natural gas industry has helped bring rapid and sophisticated human development to poor areas. Generally, the oil industry has brought urban growth and welfare improvements to simple herding, gathering and hunting economies.

A price to pay

The industry, however, has also abused deserts. For many
years, natural gas, a by-product of oil, was simply burnt off as
a waste product, an action that generates atmospheric
pollution. In the North American oil fields, thousands of
hectares (acres) of land are exploited by surface pumps and
the surface is scarred with spoil. The low value of deserts
leads to the use of unsuitable or dangerous technologies for
storage and pipeline construction, while local flora and fauna
are disregarded. Only in the United States has any form of
conservation been attempted. The worst pollution occurs
with disasters. In Kuwait, when Iraqi troops destroyed
equipment, including well-heads, in oil fields during the
1990–91 Gulf Crisis, millions of barrels of oil were spilt onto
the land and into the waters of the Persian Gulf. The
environmental damage will take years to repair.

Perhaps equally harmful have been the activities of
governments with oil revenues to spend. Belief in the need to
stimulate "development" has led to rapid growth of
agricultural and industrial projects and their associated towns
and infrastructures – all paid for by oil. In many oil-exporting
countries, notably Saudi Arabia, huge petrochemical
industries have been set up, bringing to the desert unforeseen
problems, particularly the disposal of industrial waste. But
the natural resources in most desert states are unable to
support sustained heavy industrial development and there is a
real chance that for all the long-term negative environmental
and ecological impact of modernization, human benefits will
only be short-lived.

Above *Syrian riggers examine a
drill bit on the Omar oil-field. In the
background, a newly assembled
drilling rig awaits their decision.
Drilling in remote areas means that
every action must be carefully
planned. Although surveys (and
existing wells) may indicate the
presence of oil beneath the ground,
there is no substitute for "sharp-
end" experience – the colour,
texture, smell and taste of the
material on the drill bit can all
give vital clues.*

MINING AND MINERALS

Most of the industry in deserts is based on extracting hydrocarbons and minerals. Apart from a small number of enterprises, such as tourism and film-making, that can gain some advantage from the climate, deserts offer few other positive attractions to industrialization.

Land is cheap in deserts, so large-scale mineral extraction operations can be undertaken using huge areas for mining, marshalling and depositing of ores and spoil. Exploration and survey can be readily carried out in desert regions where the soil is thin and vegetation sparse. The visual pollution arising from these activities has carried little stigma until recently. Pollution of the atmosphere, the land and water supplies has been widespread, and with few exceptions extractive industries in deserts have been exploitative and temporary.

The largest single extractive industry in deserts is for building materials. Most activities are small scale, producing materials such as road stone, lime, chalk, gypsum and building sand. This industry generally serves local markets because transportation costs of low-grade minerals are high in relation to the minerals' value.

Diamond mining in Namibia

Namibia's richest areas for diamonds are on the Diamond Coast, where the Namib Desert meets the Atlantic Ocean. The diamonds were formed in what is now South Africa, and were washed to the delta of the Orange River. From here coastal currents carried them north and deposited them on beaches along with other sediments. The beaches were buried as sedimentation continued.

Modern extraction methods involve huge open-cast pits. Massive earthen walls are built to hold back the sea, and excavation takes place at the level of the bedrock, as much as 20 metres (65 feet) below sea-level. The diamond mines leave mountains of discarded rock, and when abandoned the workings flood with seawater to leave a huge chequered pattern along the coast.

Inset left *Two Bolivian miners take a rest from pushing an ore truck across a scarred mountainside. A miners' church stands on the high place behind them. In the background an overhead bucket-line carries the ore over a ridge. Metal-mining is the backbone of the Bolivian economy; the country has large reserves of tin, copper, lead, antimony, zinc and tungsten ores. Extraction costs are high, however, largely due to the costs of transportation across such difficult terrain.*

Left *Its terraced contours highlighted by snow, the crater of the world's deepest open-pit mine competes in grandeur with the mountains near Salt Lake City, Utah, USA. The bottom of the Bingham Canyon copper mine, now covered by a small lake, lies more than 800 m (2,500 ft) below the level of the original land surface.*

Minerals of the desert

Much of the other mining that takes place in deserts is not directly related to the desert itself but to what lies under it. Some mineral are, however, related to the conditions prevailing in some deserts. The world's biggest deposits of phosphates, for example, are found in desert environments – particularly the Sahara and Jordanian deserts – and these are traded internationally on a large scale.

One of the world's hottest desert areas, the Dead Sea, is the centre of two large potash industries, one In Israel and the other, just across the border, in Jordan. Both extract potash from the salty water on the shores of the Dead Sea, where falling water levels in the early 1990s caused concern to both parties. The potash is used in domestic industries and are exported, mainly as fertilizers.

In the Western Desert of Iraq, at sites such as Al-Qa'im, phosphates and natural sulphur are exploited for the chemical and petrochemical industries. Elsewhere, sulphur is produced as a by-product of crude oil. Other minerals, including common salt, are mined in many deserts and provide the natural resource for local chemical industries.

Metals and precious stones

Large-scale reserves of metallic ores are found in many deserts, some of which are important to world trade. Most copper, for example, comes from the fringes of the Atacama in South America and from Sar Cheshmeh in the high salt deserts of the Iranian Zagros Mountains. Considerable reserves of iron ore exist in the Sahara in Mauritania and southwestern Algeria, and in the Dasht-e Kavir Desert in Iran. Gold, silver and precious stones are commonly associated with the mines in Australia, as well as in the Kalahari and North American deserts. Since the start of the nuclear age, uranium mining has become increasingly important. Some desert regions of Australia, the United States and Namibia hold large reserves of uranium ore.

DESERTIFICATION

If the deserts really were expanding, the threat would be critical. They could enlarge if the climate changed or the desert edge were gravely mishandled. Both have happened and will unquestionably happen again. In the course of the past 12,000 years, there have been times when climate change has shrouded millions of now vegetated square kilometres with desert. Human activity has also created deserts; in Iraq, salt injected from irrigation schemes dating from before the time of Christ has endured in many millions of hectares of soil. But is desertification happening now?

At the time of the great Sahel drought of the late 1960s and early 1970s, some scientists asserted that the world was undergoing a real climatic change. There are also those today who believe that global warming may lead to the dessication of large parts of the world. On the other hand, the climate of the desert is fickle, and there are few meteorological stations in place to measure such a shift as it might affect deserts.

Meteorologists now concede, however, that rainfall in the West African Sahel was below the 30-year average in the late 1910s and in the 1930s, and that the period 1970–90 was dry

by the standards of previous years. The cruel and rampant droughts that occurred in the Sahel and the Horn of Africa during this period are ample evidence. It is uncertain whether this long dry period has ended, or whether the better rains that fell were themselves abnormal. For people living on the desert edge, drought will always be a fact of life, but we should not confuse short-term fluctuations with long-term climate change of the sort that could lead to major changes in the extents of the world's deserts.

For some, desertification includes the loss of soil fertility brought about by persistent cropping of the same land, and even so-called "green desertification" – the invasion of pastures by trees and shrubs (reducing their productivity for grazing cattle). These confusions of meaning have led to false moves in the fight against desertification. One recent plan, financed with Japanese aid, called for the planting of a massive "green belt" from Mali to Chad, when most experts realize that the real problems occur some distance further south. To avoid such confusion, many scientists prefer to call these persistent problems "land degradation".

Salinization, or the build-up of salt in the soil and in the water, ranks among the most serious problems in arid lands. When irrigation water evaporates, salts are unavoidably concentrated at the surface, or creep back into the "return flow" that reaches rivers downstream. Thus an upstream irrigation scheme salinizes the water for those downstream. There is little that can be done to prevent this process in the future, apart from maintaining a good flow of fresh water through the soil, and the provision of expensive special conduits to take away the salty return flow.

Soil erosion is the second greatest problem in dry lands. Erosion occurs as a result of clearing the soil for cultivation. Erosion is probably worst in highland areas on the desert edge, such as northern Ethiopia.

Threats – perceived or real?

For some scientists, intensive grazing is a further contributor to land degradation. This view holds that grazing animals could severely and permanently damage the vegetation of the desert edge. It has led to calls for the reduction of animal numbers to a point known as the "carrying capacity" of the land. However, for most pastoral peoples, stock means survival. These people are well aware that in good years they can maintain numbers well above any official "carrying capacity", and return to these numbers after bad years. Much of the degradation perceived by early scientists as being caused by grazers was in fact caused by drought. After one drought, pastoralists can seldom build stock numbers up to levels that can damage pastures before the next drought intrudes. The ecosystems of the desert edge are adapted to drought, and can recover remarkably quickly.

There is a growing number of people who dismiss the concept of desertification. The danger this raises is that the very real problems of managing dry lands will be overlooked, especially the perennial problem of drought and the looming threat of climate change.

Left *Cattle being watered in Niger, West Africa. The congregation of cattle each day has trampled away the sparse cover of grass for many yards round the well, but the tree may even benefit when the competing grass has gone. The trampled circle is a necessary sacrifice for this mode of cattle-keeping. Beyond the bare area the damage is probably much less.*

Below *Villagers strip firewood from a living tree in Mali, West Africa, leaving green foliage lying on the ground. Fuelwood is still the main source of domestic energy for millions of people, and is becoming very scarce in many places. Killing a tree leads to the destruction not only of the tree itself but of a sheltered micro-habitat that might have supported other plants.*

Right *Opinions vary greatly as to which areas of the world, if any, are at serious risk of desertification. The map here shows the results of one study into the question, and indicates the outer limits of areas that may be at risk in the next 30 or 40 years. The darker shading shows areas defined in this book as being arid. This map demonstrates the most pessimistic end of the spectrum of opinion; indeed, many scientists do not consider that deserts are likely to expand significantly at all, except perhaps in certain limited areas. The study that produced these data reflected, however, the concern many people feel at the possible effects of land degradation and climatic change.*

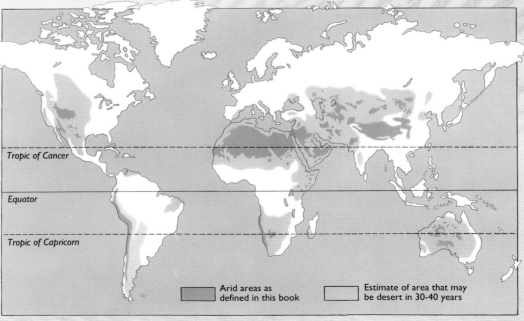

Tropic of Cancer

Equator

Tropic of Capricorn

Arid areas as defined in this book

Estimate of area that may be desert in 30-40 years

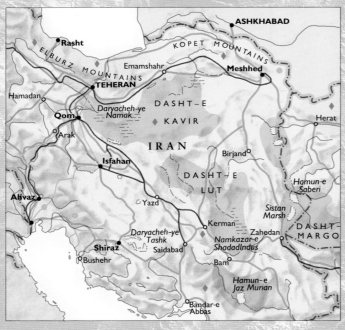

THE ATLAS

From the great sand seas of the Sahara to the foggy coastal plains of Peru, from the broken rock of the cold Gobi to the searing volcanic heat of the Danakil, the variety of desert environments is greater even than the number of deserts themselves. The Atlas explores, region by region, the deserts of the world. It looks at the physical characteristics of the deserts and the great forces of nature that formed them. It describes the lives of the peoples that inhabit the deserts and their unique and extraordinary cultures. The resources of deserts – both living and mineral – and how they are exploited are discussed, together with the threats to the ecologies and the peoples of the deserts and their margins.

Above *Maps show the locations and extents of deserts.*
Right *Satellite images give a new perspective on desert landforms.*

THE WORLD'S DESERTS

Deserts comprise over 30 per cent of the Earth's land surface. They cover very high proportions of the continents of Africa and Australia and almost all of southwest Asia and Central Asia, as well as all of the southwestern states of the United States and much of northern Mexico. There are no deserts in Europe and only small parts of South America are desert (although these are particularly extreme in character).

The extent of desert regions is not fixed. Every year weather conditions vary. Periods of dry years may occur together and establish local patterns of aridity which may prevent the regeneration of natural vegetation and so affect the extents of deserts.

Deserts and remote sensing

The Atlas section of this book combines maps of the world's deserts with images taken by remote sensing satellites orbiting the Earth. Since the 1960s, the Earth has been scanned daily by satellites carrying cameras and other sensing devices. Since the 1970s there have also been geostationary platforms with sensors scanning hemispheres of the globe every half-hour. The information captured by some of these sensors provides images closely resembling the true colour photographs with which the eye is very familiar. However, most of the information recorded is from parts of the electromagnetic spectrum outside the visible wavelengths.

Unfortunately, there is no rational system for converting the huge volume of remotely sensed data into visible colours. In the visible wavelengths, objects and surfaces which appear red in colour are reflecting the red light that falls on them while at the same time absorbing the ultraviolet, blue, green and yellow parts of the spectrum. Vegetation, on the other hand, appears green because it absorbs red and other wavelengths and reflects green light. Vegetation also strongly reflects infrared light, but this light is invisible to the eye. Because infrared light is recorded by remote sensing systems and conveys information, it is important to be able to show this information on photographic images.

It is conventional to show the infrared reflection in red. Hence on many of the pictures in the Atlas where heavily vegetated areas are present, such vegetation appears red. Heavily cropped areas, such as the irrigated delta and valley of the Nile, thus show in red. In the wet season, marginal areas, such as the Sahel region of Africa, will show as versions of red according to the degree of vegetation cover present in that particular year. Satellite imagery is very useful for recording land surface changes during a season, and from year to year, and will be an important source of information on the controversial issue of desertification.

The spectacular imagery contained in the section that follows exemplifies the capability of remote sensing to provide information about vast areas. However, satellite images can record other information than the presence and absence of vegetation and the varying geology of desert regions. Satellite-borne sensors can also record temperature and soil moisture, and meteorological satellites record the weather systems that determine both the terrestrial and ocean environments, including those of the world's deserts.

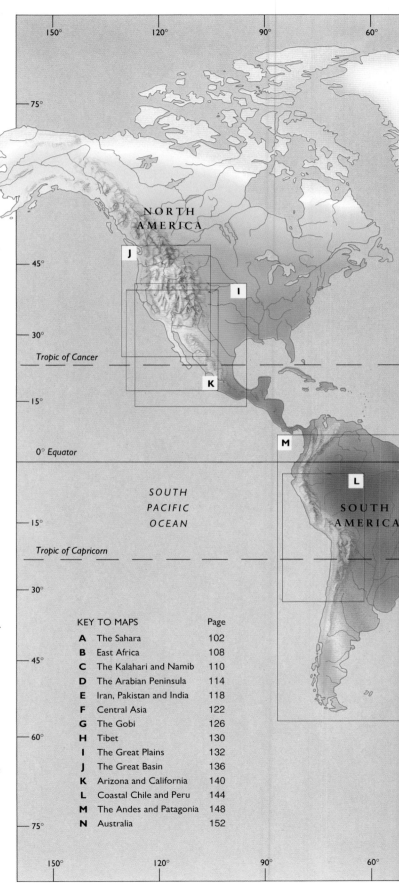

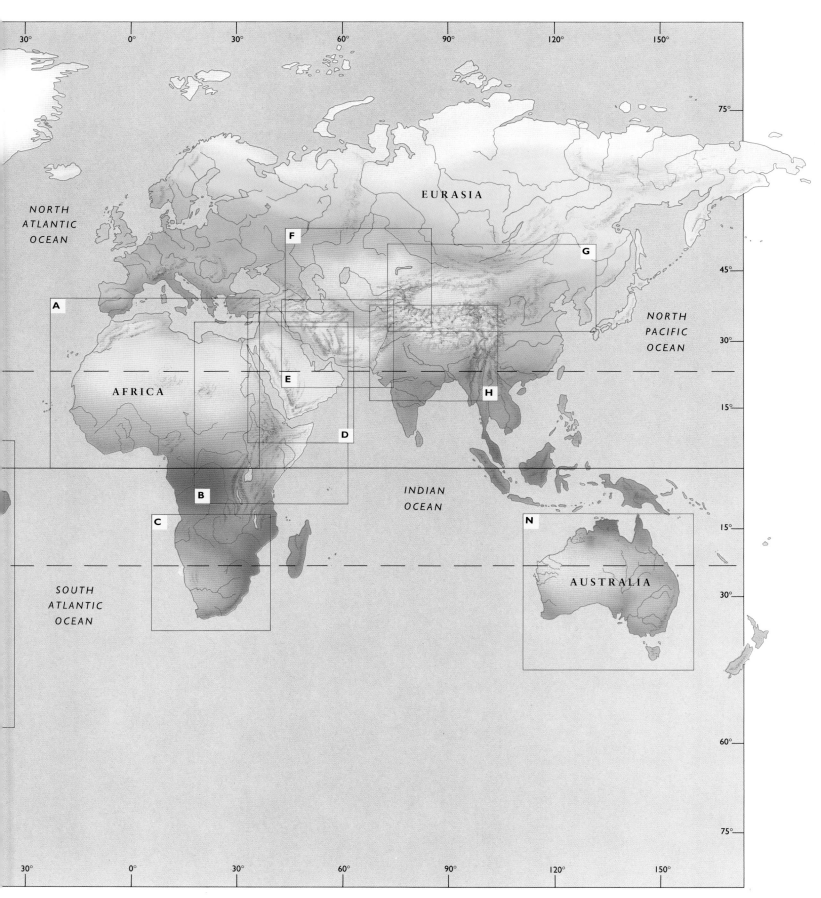

NORTH
ATLANTIC
OCEAN

EURASIA

NORTH
PACIFIC
OCEAN

AFRICA

INDIAN
OCEAN

SOUTH
ATLANTIC
OCEAN

AUSTRALIA

A
B
C
D
E
F
G
H
N

30° 0° 30° 60° 90° 120° 150°

75°
45°
30°
15°
15°
30°
60°
75°

THE SAHARA DESERT

The Sahara is the world's largest desert, a wilderness of stony plains, mountains, high rocky plateaus and barren immense sand seas, stretching from the Red Sea in the east to the Atlantic Ocean in the west, and from the southern shore of the Mediterranean to about 10° latitude. The Sahara covers 9 million square kilometres (3½ million square miles), and crosses the boundaries of 11 states.

Such a huge area inevitably includes a range of climates. The great, hyper-arid core desert of the centre has immensely hot summers, when temperatures can reach well over 55°C (130°F) in the shade. In places like Kufra, in Libya, or Toudenni, in Mali, it may not rain for years on end. But on the northern fringes, even of this core desert, it is not uncommon to see hoar frost on the dunes on a winter morning. In these parts there is virtually no vegetation, except along some of the larger wadis, where a few Acacia trees may survive against the odds, or in the very few oases. Moving north or south, the rainfall increases, and low scrub, sparse grass and then trees appear. In the south, a new hazard intrudes: the strong *harmattan* wind blows in winter, obscuring the Sun with dust. In contrast, the summits of the Aïr and Tibesti are quite cool, especially in winter, and support a few bushes of Mediterranean flora.

The landscape

Although created on an eroded block of very ancient rock, the Sahara contains an astonishing variety of landscapes. In the centre are the two massive volcanic blocks of Aïr and Tibesti. The volcanic cones are still preserved in the Tibesti, but all that remains in the Aïr are the great pillars of lava plugs. The volcanic rocks rest on ancient sandstone plateaus, which connect the Aïr and Tibesti massifs and spread out through the central Sahara.

On the fringes of these plateaus there are broad, stony, nearly featureless plains, known as *serir*. The two largest are the Calanscio Serir in Libya and the Tanzrouft in Algeria and Mali (crossed each year by the Paris-Dakkar rally). These are as big as France, and can take days to drive aross.

Sand seas cover one-third of the Sahara. In the Issouane n Arrararene in Algeria, some dunes reach 122 metres (400 feet) in height. The greatest sand seas, or *ergs*, lie in the north, notably the Great Eastern and Great Western sand seas in Algeria, the Idehan Mourzouk in Libya and the great sand seas of Egypt's Western Desert. The Great Eastern Erg covers 192,000 square kilometres (74,000 square miles). Between the shifting dunes, long corridors of rocky or coarse sandy soil carry the caravan routes, and preserve the remains of lakes that filled the hollows in wetter times.

Human settlement

In prehistoric times, the Sahara was much better watered. The evidence includes fish-hooks in areas where rainfall is now seldom even an annual event. During the 4000s BC, the Sahara began to turn arid again. The re-establishment of the desert caused the north and south to develop separately, and two migrations of Berber and Arab peoples into the northern areas increased this division.

Desert

Semi-Arid

Main roads

Other routes

Railways

Rivers

Seasonal rivers

Seasonal lakes

Mining/mineral exploitation

Oil and gas fields

Oil pipelines

Oasis

Large cities bordering desert

Important desert towns/settlements

International borders

Disputed borders

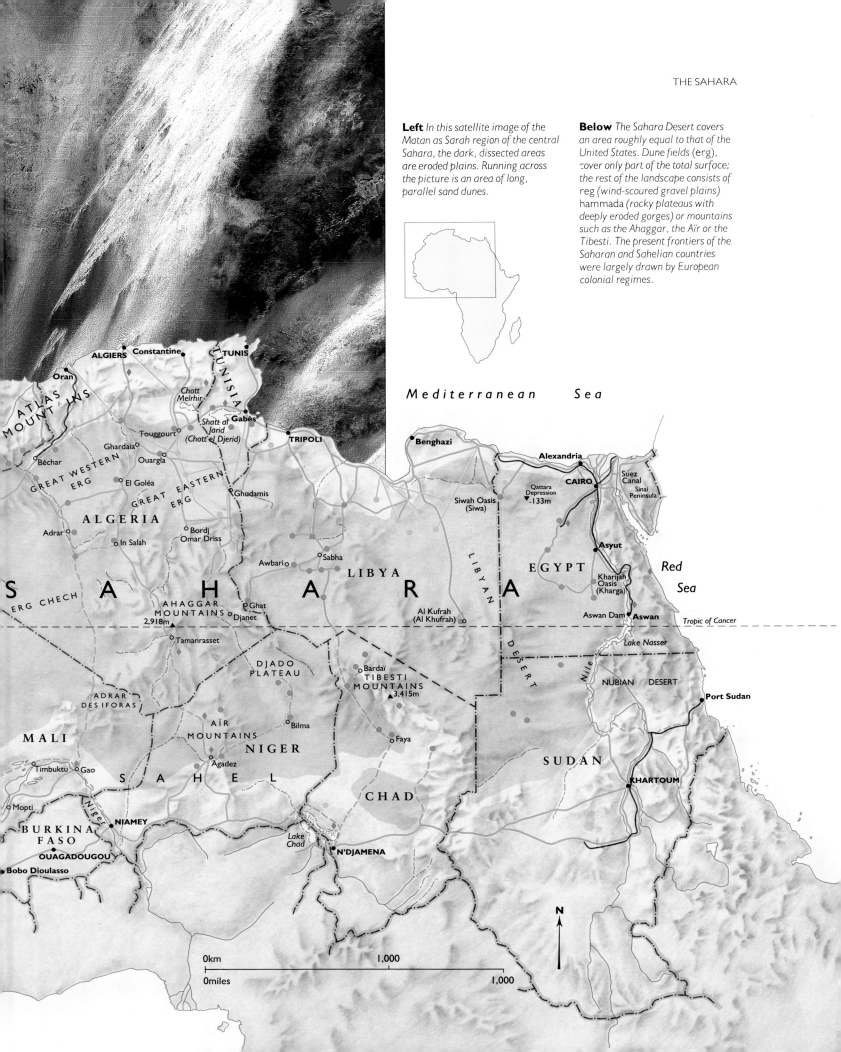

Left *In this satellite image of the Matan as Sarah region of the central Sahara, the dark, dissected areas are eroded plains. Running across the picture is an area of long, parallel sand dunes.*

Below *The Sahara Desert covers an area roughly equal to that of the United States. Dune fields (erg), cover only part of the total surface; the rest of the landscape consists of reg (wind-scoured gravel plains) hammada (rocky plateaus with deeply eroded gorges) or mountains such as the Ahaggar, the Aïr or the Tibesti. The present frontiers of the Saharan and Sahelian countries were largely drawn by European colonial regimes.*

Mediterranean Sea

ALGIERS Constantine TUNIS
Oran TUNISIA
Chott Melrhir
ATLAS MOUNTAINS
Shatt al Jarid (Chott el Djerid) Gabès
Touggourt TRIPOLI
Ghardaia
Béchar Ouargla
GREAT WESTERN ERG
El Goléa
GREAT EASTERN ERG
Ghudamis
ALGERIA
Adrar
Bordj Omar Driss
In Salah
Awbari Sabha
LIBYA
Benghazi
Alexandria
CAIRO
Suez Canal
Sinai Peninsula
Qattara Depression -133m
Siwah Oasis (Siwa)
LIBYAN
EGYPT
Asyut
Red Sea
Kharijah Oasis (Kharga)
SAHARA
ERG CHECH
AHAGGAR MOUNTAINS 2,918m Ghat
Djanet
Al Kufrah (Al Khufrah)
DESERT
Aswan Dam Aswan
Tropic of Cancer
Tamanrasset
Lake Nasser
DJADO PLATEAU
Bardaï
TIBESTI MOUNTAINS 3,415m
NUBIAN DESERT
Port Sudan
ADRAR DES IFORAS
AÏR MOUNTAINS
Bilma
Nile
MALI
NIGER
Faya
SUDAN
Timbuktu Gao
Agadez
SAHEL
KHARTOUM
Mopti
CHAD
BURKINA FASO
Niger
NIAMEY
Lake Chad
N'DJAMENA
OUAGADOUGOU
Bobo Dioulasso

N

0km 1,000
0miles 1,000

SAHARAN RESOURCES

The Sahara is a hostile environment. Rainfall is too scarce to support crops, and even livestock rearing is not generally a viable activity. This barren environment can barely support wildlife, livestock and people, and the flora and fauna that do exist have had to adapt dramatically in order to survive. Human communities of the Sahara throughout history have developed a range of remarkable and inventive strategies in their struggle for survival.

The Sahara is crossed by two major river systems – the Nile and the Niger – and a number of smaller ones, such as the Lake Chad system and the Oued Saoura in Algeria. These invaluable water resources support the majority of the human population of the desert. Out of the 55 million people who live in the Sahara region, the river systems provide livelihoods for about 43 million of them. Although this amounts to less than one per cent of the world's population, the desert they inhabit comprises about eight per cent of the Earth's land area. The segment of the Saharan population that relies on the resources of the desert itself, rather than the river systems, amounts to only about 12 million people, or a mere 0.2 per cent of the world's population. This makes the Sahara one of the emptiest regions anywhere on Earth.

An infinite resource?

With surface water so scarce and unreliable, groundwater has always been enormously significant. The groundwater of the Sahara is often of a very high quality and suitable both for agricultural and domestic uses. Over half of the exploitable water is ancient water, which fell as rain many thousands of years ago. These water resources are finite, and are usually referred to as "fossil" water. Most of the groundwater of the northwestern Sahara, particularly in Libya and southwest Egypt, consists of fossil water. The major aquifer that underlies Algeria and southern Tunisia, the Continental Intercalaire, is recharged from the meagre rains that fall on, and gradually filter down into, this vast reserve.

The most spectacular development of Saharan groundwater retrieval was carried out in Libya during the 1970s and the early 1980s. At this time, the water was used mainly in the Kufrah area and in Fezzan. However, the schemes based on these remote water resources were not economically or socially successful. Accordingly, development emphasis shifted in the late 1980s and early 1990s with the construction of massive pipelines to convey water from the aquifers to the Mediterranean coastal region.

Above An oil pipeline makes a sudden change of direction in the Sahara Desert. The concrete blocks that originally supported the pipeline have been displaced by the shifting sands of the surface, and much of the pipeline is now supported by a ridge of sand. In the background, a second pipeline is being laid parallel to the first.

Right The life-giving Nile River in southern Egypt. Two goatherds try their luck at fishing, while their animals graze on the bank. Behind them lie green fields irrigated with water, occasionally still lifted from the river by human muscle power. Behind the fields are tall palm groves, which provide dates, oil and useful fibres. In the far distance lies the arid escarpment that marks the edge of the Nile flood plain. Beyond lies bare open desert.

Right *Workers at a Moroccan phosphate mine shelter from the Sun in the shade of a giant conveyer belt that carries part-processed phosphate rock to the coast for export. Although phosphate extraction does not require sophisticated technology, it does require the large-scale application of expensive heavy engineering in order to be commercial. The additional costs of atmospheric pollution generated by the processing plant in the background can easily be imagined.*

Mineral resources

In economic terms, minerals have proved to be the Sahara's most important natural resource. Libya and Algeria have benefited greatly from their oil and natural gas resources. Both countries derive over 90 per cent of their export earnings from these hydrocarbons. Tunisia and Egypt have also benefited from oil, but to a much smaller extent.

Other mineral resources have been, and remain, major contributors to the economies of some Saharan countries. Phosphates, for example, are Tunisia's major export, and while Morocco, Mauritania and Western Sahara are also major phosphate producers, the stability of these exports has been disrupted by the insecurity and military activity of their border zone. Iron ore resources are also significant, especially in Libya, and have led to processing and manufacture.

Another of the Sahara's major natural resources, sunlight, is as yet undeveloped because of the primitive state of the technology for solar energy conversion. Huge amounts of sunshine fall on the region throughout the year, and the Sahara's vast land area gives it the greatest potential for solar energy development of all the world's deserts. However, with technology in its current state, huge tracts of the desert would have to be covered by solar energy receptors to provide even a meagre supply of electricity. This would, unfortunately, involve unacceptable levels of investment. But the Sahara's comparative advantage for solar energy generation could transform the future potential of the region for economic activity. Cheap and accessible solar energy would enable a wide variety of urban economic activities to be sustainable. Solar energy could even be significant in mitigating the water shortages, since it could provide the power for desalination schemes. In turn, desalination would raise the quality of the widely available, local brackish water to that needed for agriculture and other economic uses.

THE TUAREG

When the Arab armies moved across North Africa during the AD 600s and 700s, they replaced the ruling Berber population. While much of the Berber population was and remains sedentary, some were nomadic. The most famous of these are the Tuareg, whose indigo-robed male warriors provide one of the Sahara's most familiar images. After the first Arab conquests, the Tuareg were initially reluctant to adopt the new Islamic religion, but over the following centuries this reluctance was replaced by a devout adherence to Islam. Tuareg men must wear a veil, known as the *tagilmust* (a long strip of cotton, often dyed blue). Women wear a smaller veil, which covers only the mouth.

The Tuareg number around one million, and are divided into seven groups or confederations of tribes. The Tuareg homeland lies in six states: Mali, Burkina Faso, Mauritania, Niger, Algeria and Libya. For centuries they have followed a nomadic or semi-nomadic way of life, based on the raising of camels, goats, sheep and cattle.

The Tuareg speak dialects of Tamahaq, itself a form of Senhadjan Berber. Dialects vary according to the tribal confederation. The Ahaggar Tuareg confederation is itself broadly divided into three tribes: the Kela Rela, Tégéhé Millet, and the Taituq. Each tribe is headed by a clan which is distinguished as aristocrats. Traditional Tuareg society consisted of three classes: nobles, vassals and slaves. Historically, Tuareg nobles controlled the Saharan caravan routes. The vassals concentrated on herding. The slaves and former slaves, known as *harratin*, performed menial duties and guarded tribal encampments.

Tuareg culture is patriarchal, but inheritance of the chieftainship of the tribe passes through the female line, so that the successor is the eldest son of the incumbent's eldest sister. Tribal succession has always been a pragmatic affair, determined by the willingness of the tribe's clan to pay tribute to the leader, or *amenukal*.

Tuareg traditions have suffered under increasing economic and political pressures in recent years. Originally, their traditions developed in a fluid fashion, because of the difficulty in maintaining strict social divisions and hierarchies among highly mobile nomadic groups. The Ahaggar confederation has historically been forced to break up the clan and tribal divisions into smaller familial groups because the intense aridity of the area requires quick movement between water supplies and grazing areas.

In addition to the pastoralist or herding economy, Tuareg tribes have long acted as intermediaries for cross-Saharan traffic. For long periods, Tuareg nobles controlled many of the trading routes across the Sahara, supplying guides for the Arab traders of the North African coastal region.

Mechanization has eliminated many of the problems of desert transport, depriving the Tuareg of their main source of external income. In addition, the Tuareg are now suffering the most immediate effects of the long recent drought. The periodic droughts that have devastated the Sahel region since the 1960s have also drastically reduced the amount of pasture available for their herds. This situation has forced many Tuareg to settle in towns.

A threatened people

Twentieth-century Tuareg history has seen three military uprisings. The first, in the early years of the century, was easily suppressed by the French. Little further political activity was permitted until the partition of the traditional Tuareg homelands of Aïr and Azawad between the newly independent states of Mali, Burkina Faso, Mauritania, Niger, Algeria and Libya.

During the late 1970s, the Libyan leader Colonel Muammar Qadhafi began to recruit Tuareg guerrillas to fight in Chad. As a result, many Tuareg began to pursue the dream of freeing the Azawad. In Mali and Niger, groups of Tuareg fighters are now waging a guerrilla war against the governments of the countries whose boundaries cross the limits of the historic Tuareg homelands. Their demands vary from the total independence of the Aïr and Azawad to greater regional autonomy for the Tuareg areas of northern Mali and Niger. The guerrilla war, largely caused by shrinking grazing lands, looks likely to continue until the Tuareg feel that they are *imohagh* (free).

Left *Modern political boundaries dissect the central Saharan homeland of the Tuareg people. Three major upland areas act as focal centres for these people: the Aïr region of Niger, the Ahaggar region of Algeria and the Adrar des Iforas in Mali. Political boundaries also cut across the old caravan routes, hindering the movement of the remaining nomadic groups. In recent years, the increasing destruction of their traditional way of life has led many Tuareg to fight for an independent state.*

Above *The veil worn by Tuareg men protects them both from desert winds and from evil spirits that lurk in dark, lonely places. This long strip of indigo cloth is known as a tagilmust. The indigo robes worn by the Tuareg have given them the name "blue men of the desert"; however, the Tuareg refer to themselves simply as Kel Tagilmust, or "people of the veil".*

Right *The extraction of salt has been a vital economic activity for Tuareg tribes since the 1890s. Salt is carried out of the desert by caravan and traded for grain, although this activity has declined greatly since the 1960s.*

Above *For centuries, Tuareg nobles provided guides for Saharan camel caravans, which were the sole means of contact between the Mediterranean coastal regions and sub-Saharan Africa. Air transport and the construction of roads have reduced the importance of caravan traffic, and the remaining Tuareg nomads depend on livestock-herding for survival. However, drought has drastically reduced the amount and the quality of pasture that is available for their herds.*

107

EAST AFRICA

East Africa, a region sometimes loosely known as the Horn of Africa, has a group of adjoining semi-arid and arid areas that stretch through Kenya, Somalia, Ethiopia, Djibouti and Eritrea. Further to the north begin the deserts of Sudan, such as the Nubian Desert, which are essentially part of the Sahara. The dominant geographical feature of the region is the Great Rift Valley, which runs approximately southwards from the foot of the Red Sea.

The Eastern lowlands

Kenya, in the south of the East African region, lies on the Equator, but temperatures here are determined more by altitude than by latitude. The highlands to the west of the Rift Valley are generally fertile. To the east of the valley are dry plains that cover three-quarters of the country. The southern and central plains receive little and unreliable rain; most people live in the few hill areas, where the climate is slightly more temperate. In the northern parts of the Kenyan plain, which extend into Somalia and southern Ethiopia, are the Chalbi and Didi Galgalu deserts. These regions support no agriculture and have little vegetation other than small shrubs and the occasional stunted tree. The only means of survival for people in these areas is subsistence pastoralism.

The rest of the region covered by Somalia consists largely of dry savanna plains, with the exceptions of a fertile strip between the Juba and Shebelle rivers in the south and an area of highlands in the northwest. The Somali plains continue westwards into Ethiopia, becoming increasingly arid and forming the Ogaden Desert, which is bounded to the west by the highlands of the Rift Valley. The population of the Ogaden consists almost entirely of nomadic herders.

The pastoral agriculture in the eastern lowlands depends on spring rains. Failures of these rains have caused mass population movements and starvation. Long-running civil wars, particularly in Ethiopia and Somalia, have compounded the famines. At the same time, famine leads to political instability, and so a vicious circle is created.

The Danakil

In the north of Ethiopia, the sides of the Rift Valley move further apart, making the valley Y-shaped. The top ends of the "Y" extend into Djibouti to the east and Eritrea to the west. In this region, and bounded by the Red Sea to the north, is the Danakil Desert, the lowest part of which is known as the Danakil Depression. Much of the depression, which is near to the Red Sea, is about 120 metres (400 feet) below sea level. The Danakil is scorchingly hot, and the little rain it receives evaporates very quickly to leave large salt flats in depressions where water tends to gather. Active volcanoes belch out smoke and sulphurous fumes, and hot springs bubble, often with waters strangely coloured by minerals.

The only relief in this fierce landscape is where the River Awash crosses the southern part of the region, creating a narrow fertile strip. The river evaporates before reaching the sea, leaving only saltpans. The few inhabitants of the Danakil are Afar herders, known for their ferocious protection of the few resources that the region has to offer.

Below *A satellite image showing Mount Kulal (central red-coloured feature) near the shores of Lake Rudolf (left of picture) in Kenya. Lake Rudolf is situated in the Eastern Rift Valley, which runs southward from the highlands of Ethiopia. The radial drainage down the slopes of Mount Kulal is clearly visible; to the east (right), however, the transverse (left to right) pattern is overlain and obliterated by alluvial material that has been deposited by a north-south flow running down the length of the main valley (top to bottom).*

Right *The arid region of East Africa contains several deserts. The most sizeable of these are: to the north, the Danakil; in the middle of the region, the upland Ogaden Desert of Ethiopia; to the southeast are the waterless undulating plains of the Didi Galgalu, which reach to the shores of the ocean.*

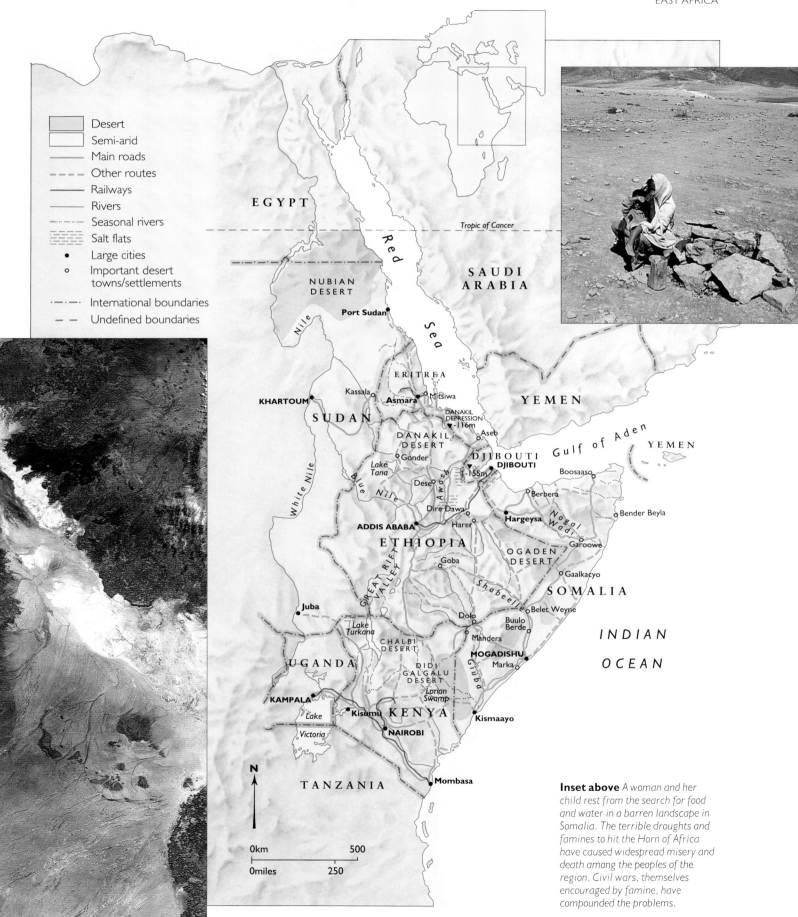

Desert
Semi-arid
Main roads
Other routes
Railways
Rivers
Seasonal rivers
Salt flats
● Large cities
○ Important desert towns/settlements
International boundaries
Undefined boundaries

EGYPT

SAUDI ARABIA

Red Sea

Tropic of Cancer

NUBIAN DESERT

Port Sudan

Nile

ERITREA

Kassala   ○ Mitsiwa

KHARTOUM   **Asmara**

YEMEN

DANAKIL DEPRESSION -116m

SUDAN

Aseb

DANAKIL DESERT

White Nile

○ Gonder

Lake Tana

○ Dese

DJIBOUTI
DJIBOUTI
-155m

Gulf of Aden

YEMEN

Boosaaso ○

Blue Nile

Awash

Dire Dawa

Berbera ○

Hargeysa

Nogal Wadi

○ Bender Beyla

ADDIS ABABA   Harer

ETHIOPIA

○ Garoowe

OGADEN DESERT

○ Goba

Shabeelle

○ Gaalkacyo

SOMALIA

GREAT RIFT VALLEY

Juba

Dolo ○

Belet Weyne ○

Buulo Berde ○

Ghuba

INDIAN OCEAN

Lake Turkana

○ Mandera

CHALBI DESERT

MOGADISHU

Marka ○

UGANDA

DIDI GALGALU DESERT

Lorian Swamp

KAMPALA

Lake Victoria

Kisumu   KENYA

Kismaayo ○

NAIROBI

N

Mombasa

TANZANIA

0km   500
0miles   250

Inset above *A woman and her child rest from the search for food and water in a barren landscape in Somalia. The terrible droughts and famines to hit the Horn of Africa have caused widespread misery and death among the peoples of the region. Civil wars, themselves encouraged by famine, have compounded the problems.*

109

THE KALAHARI AND NAMIB

The Kalahari and Namib are separated by a ridge of hills and mountains running from north to south through central Namibia, forming part of southern Africa's Great Escarpment. To the west of this ridge, the Namib drops in altitude from 800 metres (2,600 feet) to sea level. To the east, the great plain of the Kalahari extends into Botswana at an average elevation of some 950 metres (3,100 feet). If the Kalahari seems monotonous, then the Namib is notable for the diversity of its empty environments.

The Kalahari occupies some 450,000 square kilometres (174,000 square miles), covering most of Botswana and parts of Namibia and South Africa. Its dominant characteristic is its extensive sand cover – one of the largest unbroken stretches on Earth. Despite the large areas of sand dunes in the Kalahari, most of the desert supports various savanna vegetation communities. Mean annual rainfall ranges from 150 millimetres (6 inches) in the southwest, to 500 millimetres (20 inches) in northern Botswana. However, there is a notable lack of surface water. There are few perennial rivers, but seasonal rainfall accumulates in natural waterholes, known as pans. The largest of these are the Etosha Pan in northern Namibia and the Makgadikgadi Pans in northeast Botswana. At the desert's northern limit, the Okavango Delta and Chobe River offer permanent water sources. These vital wetlands form an important destination for migration by the great herds of antelope that, until recently, were an integral part of the Kalahari ecosystem.

The arid coastal strip

In marked contrast to the Kalahari, the Namib is the classic example of a hyper-arid, sub-tropical, coastal desert where ocean fog provides the main source of moisture. It forms a strip some 30–140 kilometres (20–90 miles) in width. The Orange River is usually considered to be its southern limit, with the Cunene River as its northern boundary.

The Namib exists as a consequence of very low rainfall in a region of sub-tropical high pressure. Aridity is intensified by the cold Benguela Current, which flows north along the Atlantic coast from Antarctica. The current reduces the occurrence of rainfall, but also creates the early morning fogs. Annual rainfall at Gobabeb, about 60 kilometres (40 miles) inland, is around 25 millimetres (1 inch), but it also receives around 30 millimetres ($1\frac{1}{4}$ inches) of precipitation from fog, on an average of 37 days per year. At the coast, there are 60–80 foggy days each year. Temperatures in the Namib are moderated by the fogs and by sea breezes, so that the highest temperature recorded at Walvis Bay is only 35°C (95°F). These conditions make this desert one of the world's kindest hyper-arid environments.

South of the Kuiseb River, there is an area of huge linear and star-shaped dunes – among the largest in the world. For most of the year, the Kuiseb is dry, but it flows often enough to prevent the northward movement of the dunes. The river does not flow every year and has not reached the sea since 1933. North of the Kuiseb, there are gravel plains with isolated mountains, known as inselbergs ("island mountains"), and some more massive mountain ranges.

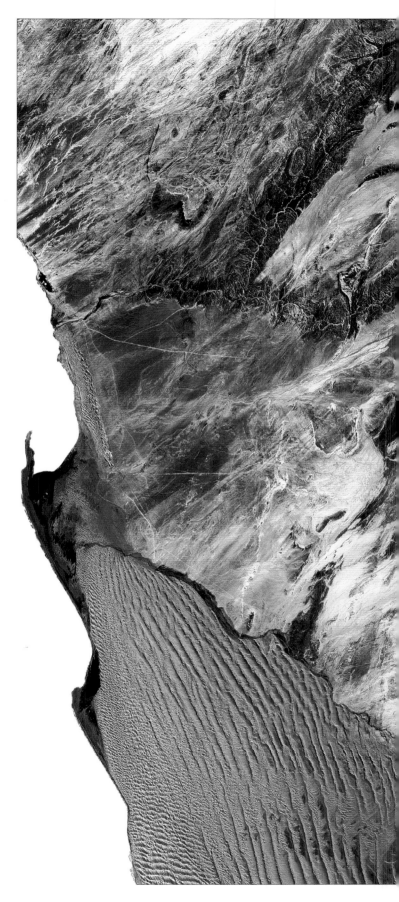

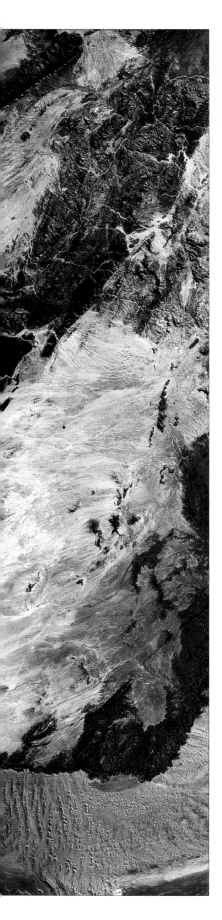

Above *The Kalahari Desert, located in the centre of southern Africa, reaches southward to the Orange River on the southern border of Namibia. A part of the interior tablelands, the Kalahari has an average elevation of around 1,000 m (3,000 ft). The Namib Desert occupies a narrow strip along the southwestern coast between the High Karoo and the Skeleton Coast.*

Left *A satellite photograph of the northern edge of the Namib Desert in southern Africa. Halted dramatically along the banks of the seasonal Kuiseb River, the linear dunes of the inner desert have been diverted westward into the Atlantic Ocean. Desert sand deposited in the shallows has formed offshore spits, altering the shape of the coastline.*

- ▒ Desert
- ░ Semi-arid
- — Main roads
- --- Other routes
- — Railways
- — Rivers
- —·—·— Seasonal rivers
- ⬭ Seasonal lakes
- ▤ Salt flats
- ◆ Mining/mineral exploration
- ▨ Irrigated zones
- ▢ National parks/ game reserves
- ● Large cities
- ○ Important desert towns/settlements
- —·—·— International boundaries

FARMING AND MINING

Until the late 1960s, the Kalahari appeared to offer little potential other than to groups of hunting and gathering San (or Bushmen). The San traditionally lived in groups of around 25, deriving their food from berries, melons, nuts, roots and seeds, supplemented by wild game. Today, there are thought to be less than 2,000 San living in the traditional manner. However, other economic activities in the Kalahari have undergone great expansion in recent years.

Cattle ranching in the Kalahari developed rapidly in the 1970s. Previously, there had been some local attempts at ranching, notably in the northern Cape Province of South Africa and the Ghanzi area. Expansion was made possible by the widespread sinking of boreholes to tap groundwater, government planning and market support from the European Community. At the same time, the sands of the Kalahari started to reveal other resources, notably diamonds. There are now three large diamond mines in Botswana, and significant deposits of coal, cobalt, copper and nickel have been found. The Makgadikgadi basin, an ancient dried lake, also now supports southern Africa's largest soda ash plant.

The rapid development of the Kalahari, which turned Botswana from Africa's poorest nation upon independence in 1966 to one of the wealthiest, has raised many environmental issues. The cattle industry has been affected by serious droughts, and the need for pasture has raised the spectre of land degradation through overgrazing. Wildlife populations have been dramatically reduced through a combination of drought, competition for land and the presence of a network of fences, constructed to control the spread of foot-and-mouth disease, which has disrupted seasonal game migration. Yet the human population is still extremely sparse and 17 per cent of Botswana – virtually all of it in the Kalahari – designated as game reserves and national parks. Wildlife management areas are also being created, where traditional methods will utilize game rather than livestock.

Exploitation of the Namib

The Namib lies in the recently independent nation of Namibia. Formerly known as South-West Africa, the territory was colonized by Germany in the late 1800s, but was controlled by South Africa from 1916 until 1990. South Africa retains control of the port of Walvis Bay.

Because the coastal strip of the Namib combines hyper-aridity with shifting sands, there has been little attempt to settle there. However, mining companies have exploited Namibia's mineral wealth, first diamonds and more recently uranium. For many years, large parts of the coastal strip were closed areas, controlled exclusively by the mining companies.

Around the inland fringe of the Namib, particularly in the northeast, there are areas where land degradation is becoming a problem. In semi-arid Namibia, annual rainfall totals can be as high as 600 millimetres (24 inches) – higher than some parts of northern Europe – but this rainfall is extremely seasonal, and is combined with very high rates of evaporation. It is in these marginal lands at the periphery of the hyper-arid desert cores where most people settle, and encounter the greatest environmental problems. Until the late 1970s, the

South African government pursued a policy of resettling the country's black population in "homelands" on the edges of both the Namib and the Kalahari. Concentration of population in these marginal areas led to familiar problems of overuse of land and of deforestation in the search for fuel, resulting in impoverishment of the soil and vegetation.

Wildlife of the Namib

The Namib is remarkable for its great diversity of animal and plant species. Most notable is the beetle population, many species of which are found only in the Namib and thrive on the moisture provided by the fogs. Around 50 species of tenebrionid beetle rely on this moisture. Some dig trenches to catch moisture, others catch condensation on their bodies. This requires the beetle to be active often at the coldest part of the day – not easy for a cold-blooded creature. The beetle population provides food for lizards, geckos, snakes, gerbils and golden moles, which then become the prey of larger animals. The sparse vegetation supports ostriches and antelope species such as springbok and oryx. Within the river valleys, carnivores such as hyena and jackal may be found.

Left *The Nossob River valley in Botswana, a well-watered region by Kalahari standards. The presence of water is confirmed by the the relatively large number of trees. However, the skeletons of dead trees standing among the living provide evidence of the progressive desiccation and degradation of the once fertile areas around the fringes of the Kalahari. In many places, a combination of drought and overgrazing have led to a massive increase in soil erosion.*

Inset left *San tribesmen share food in the Kalahari Desert. The San have no tradition of agriculture, but are skilled at locating and identifying edible roots and seeds amid the sparse scrub vegetation in the less arid parts of the desert. Wild game provides a fairly reliable source of protein. Any meat that is not eaten fresh is cut into strips and dried to make biltong, which will then keep for some time.*

Right *San workers separate out cattle for watering on a ranch in Botswana. Although the trees give an illusion of apparent lushness, the entire landscape is as dry as the dust in the cattle compound. In the face of prolonged drought, maintaining Botswana's cattle-ranching prosperity, has called for ever more stringent management of available water resources.*

THE ARABIAN PENINSULA

The Arabian Desert extends south from the Euphrates River to the port of Aden on the Yemeni coast, and from the Jordanian port of Al Aqabah in the west to the headland of Ras al-Had on the Omani coast in the east. From this vast desert area, covering some 2.3 million square kilometres (900,000 square miles), have come some of the most important influences on the world's culture and society. The region now contains two-thirds of the world's major energy resources (as oil) and includes, not surprisingly, some of the most heavily disputed territory in the world.

Physical characteristics

The Great Rift Valley splits at the head of the Red Sea round the Sinai Peninsula. The eastern arm runs northwards between Israel and Jordan, dropping down to the lowest point on Earth's land surface at the Dead Sea and fading out in Syria, causing the mountains west of Damascus. The eastern flank of the Rift Valley rises to form uplands in western Jordan, Saudi Arabia and Yemen.

Beyond this border of highlands, the desert drops away to the plains and sand deserts of the centre, east and south, which form nine-tenths of the entire area of the peninsula. This region seldom rises above 500 metres (1,600 feet). The far west of the peninsula is also uplifted in the extraordinarily complex Oman Mountains, which were formed by the movement of an ancient sea floor over the platform of rock from which the peninsula is formed.

The platform is separated from the African continent by the rifts of the Red Sea and the Gulf of Aden. This ancient platform is covered on the eastern and central desert by layers of more recent rocks. These rocks characterize the Najd, or Central Plateau. This region is the homeland of Saudi Arabia's ruling Saud dynasty and the strict Muslim Wahhabi sect, whose *Ulama*, or senior clerics, rule the peninsula in conjunction with the Saudi royal family.

The southeastern peninsula is dominated by the intense heat and aridity of the sandy Rub al Khali, or Empty Quarter. The absence of rainfall and the scarcity of vegetation make the area uninhabitable, even by nomadic pastoralists.

Most settlement is in the eastern highlands, the Hejaz. The Islamic holy cities of Mecca and Medina lie in their shadow, and the cities of Taif and the ancient Yemeni capital of Sana were built in the highlands away from the heat of the plains.

Rainfall in Arabia is sparse and erratic, except in the western highlands. Summer temperatures in the Central Plateau can reach 49°C (120°F), while around the coastal areas this drops to about 32°C (90°F). During the winter months, November to April temperatures are a little lower, and can occasionally plummet to freezing at night.

Herding and agriculture

The Asir region in the southwest of the peninsula is the most fertile area. Here, farmers have created terraced fields on which they grow a variety of crops. The eastern lowlands, which back onto the Persian Gulf, have a number of oases which also support large agricultural settlements. Cattle, goats and sheep provide the nomads and semi-nomads with, among other things, dairy products and meat.

Right *A satellite image showing (centre) Riyadh, capital of Saudi Arabia. Riyadh is situated in Wadi Hanifa (top left to bottom right), and the drainage patterns feeding into the wadi are clearly visible. Running northeast from the city (image is oriented north) are the tracks of the Arabian Peninsula's only working railway. In several places the tracks appear to have been smothered by sand (white in image). To the west of the city, the large orange feature is a sandstone plateau that has been heavily eroded and weathered; similar outcrops lie to the east of the wadi.*

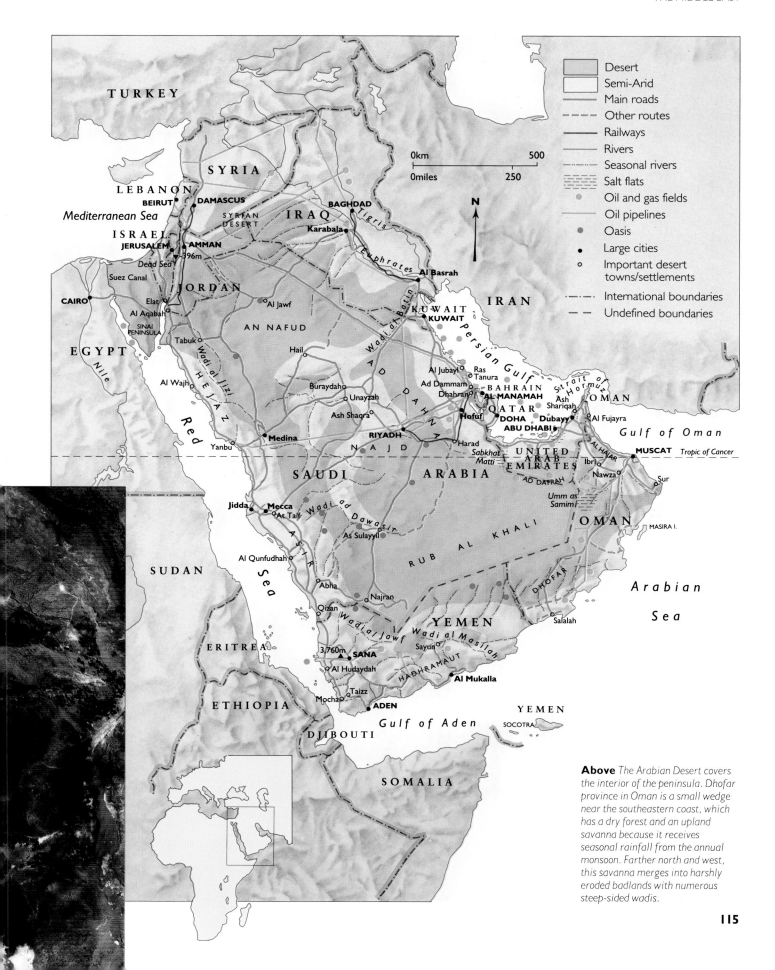

Desert
Semi-Arid
Main roads
Other routes
Railways
Rivers
Seasonal rivers
Salt flats
Oil and gas fields
Oil pipelines
Oasis
Large cities
Important desert towns/settlements
International boundaries
Undefined boundaries

0km   500
0miles   250

N

TURKEY

SYRIA

LEBANON
BEIRUT
DAMASCUS

Mediterranean Sea

SYRIAN
DESERT

IRAQ

BAGHDAD

Tigris

Karabala

Euphrates

Al Basrah

IRAN

ISRAEL
JERUSALEM
AMMAN
396m

Dead Sea

Suez Canal

Elat
Al Aqabah

CAIRO

SINAI
PENINSULA

EGYPT

Nile

JORDAN

Al Jawf

AN NAFUD

Tabuk

Hail

Wadi al Batin

KUWAIT
KUWAIT

Persian Gulf

Wadi al Hizi

Al Wajh

Buraydaho

Unayzah

Ash Shaqra

Al Jubayl
Ras
Tanura
Ad Dammam
Dhahran
Hofuf

Strait of Hormuz

BAHRAIN
AL MANAMAH
QATAR
DOHA
ABU DHABI

Ash
Shariqah

OMAN

Al Fujayra

Gulf of Oman

HEJAZ

Red

Sea

Yanbu

Medina

NAJD

RIYADH

Harad

Sabkhat
Matti

AD DAHNA

UNITED
ARAB
EMIRATES

AD DAFRAH

Dubayy

AL HAJAR

Ibri

Nawza

MUSCAT

Sur

Tropic of Cancer

SAUDI

ARABIA

Umm as
Samim

OMAN

MASIRA I.

Jidda
Mecca
At Taif

Wadi ad Dawasir

As Sulayyil

RUB AL KHALI

DHOFAR

Arabian

Sea

ASIR

Al Qunfudhah

Abha

Najran

Qizan

Wadi al Jawf

Wadi al Masilah

YEMEN

Salalah

SUDAN

3,760m
SANA

Sayun

HADHRAMAUT

Al Mukalla

ERITREA

Al Hudaydah

Taizz

Mocha

ETHIOPIA

ADEN

Gulf of Aden

YEMEN

SOCOTRA

DJIBOUTI

SOMALIA

Above The Arabian Desert covers the interior of the peninsula. Dhofar province in Oman is a small wedge near the southeastern coast, which has a dry forest and an upland savanna because it receives seasonal rainfall from the annual monsoon. Farther north and west, this savanna merges into harshly eroded badlands with numerous steep-sided wadis.

OIL IN THE MIDDLE EAST

The Arabian Peninsula lies on the rock of the Arabian plate, which is bounded by the African, Somali, and Eurasian plates. Its oldest rocks date back some 4,600 million years, but it is chalky sediments from the Mesozoic era (225–70 million years ago) which give the region its large oil reservoirs. These were formed when marine sediments were deposited in the Tethyian Sea, which lay between the prehistoric "super-continents" of Gondwanaland and Laurasia.

More than 50 per cent of the world's proven reserves of oil come from 33 "supergiant" fields of more than 5 billion barrels each; 28 of these lie in the Middle East, including nine of the ten largest. Most of the region's oil fields lie along the east coast of the peninsula and in the Persian Gulf.

Apart from oil and natural gas, the countries of the Arabian peninsula have very few other resources. There is some salt production, and there are limited quantities of zinc, lead, iron ore, copper, gold and chromite. With the exception of the southeast, low rainfall throughout the region means that agriculture is possible only through irrigation.

The impact of oil
The development of oil and natural gas has been the driving force behind the economies of the Arabian peninsula. Industry in the peninsula depends almost entirely on the refining and processing of oil, or on the cheap energy that abundant oil provides. Oil has also brought radical change to the societies of the sparsely populated desert states. A way of life built around herding and nomadism has now largely given way to a more settled, urban lifestyle.

The massive oil price rises of the 1970s brought a huge inflow of revenues to the oil states. The development, however, has a high price because it comes largely not from productive effort but from a non-renewable resource.

Although oil revenues have resulted in the establishing of some of the most modern industry and welfare structures in the world, the domestic labour force alone was too small and unskilled to accomplish this task. Foreign workers were recruited, initially from the Arab world, but increasingly from southern and eastern Asia.

Oil reserves
With proven reserves of 258 billion barrels, Saudi Arabia is the region's major producer. Oil was discovered in 1938 and development began after the Second World War. Among Saudi Arabia's most significant onshore oilfields are Al-Qatif, Berri, Khurays and Fadhili. There is also a number of extensive offshore fields.

Kuwait's first oil strike was also in 1938, and proven reserves are approximately 97 billion barrels. Kuwait's oil attracts lower prices on world markets, however, because it has a high specific gravity and a high sulphur content. Environmental considerations mean that lighter, cleaner crude oil is now increasingly preferred.

The United Arab Emirates (UAE) are major producers and have reserves of 98 billion barrels. Most of this lies in Abu Dhabi, although Dubai has a number of offshore fields. The other, smaller emirates have some reserves of gas.

Below *The refinery complex at Jubail, Saudi Arabia. Jubail is a new industrial city, planned and built by the Saudis as the showpiece of their new petrochemical industry. Despite having enormous oil reserves, the Saudi government intends to move the country away from a dependence on primary oil extraction. Billions of dollars have been invested in infrastructure projects and in industrial centres such as Jubail, which has an aluminium smelter, a steel mill, and plastics and fertilizer plants.*

Other Arab states have smaller reserves. Qatar has about 5 billion barrels, but it also has vast reserves of natural gas. Gas is increasingly popular as a clean-burning fuel, and is no longer burned off to the extent that it was. Oman's first discoveries were not made until 1964. Oil reserves are modest at just over 4 billion barrels. Yemen hopes that recent discoveries will enable it to become a significant oil exporter. The country has proven reserves of about 4 billion barrels, but there is uncertainty over the development of some of them because of disputes over the country's boundaries.

Many of the borders between the nations of the Arabian Peninsula are not clearly defined – the inhabitants of sparsely populated desert areas had no call for rigidly defined territorial limits. The discovery of oil and gas changed this situation, and has led to border disputes and even to war.

Oil pollution, already severe in the Persian Gulf, has worsened since the Iran-Iraq war of 1980–88. In 1991, Iraqi forces released an estimated 4 to 6 million barrels of oil into the Gulf during the Gulf War; this had a disastrous effect on the wildlife of the region. Inland, 150 million barrels of oil leaked into the desert from wellheads sabotaged by the Iraqis; this has already badly contaminated soils and vegetation and may seep into groundwater supplies.

Above *A Saudi survey team continues the search for oil in the sands of the Rub al Khali (the Empty Quarter). The Saudi oil industry has concentrated on employing Saudi nationals, and is no longer totally dependent on expatriate labour. A massive investment in education is now producing a new generation of highly trained and skilled workers for industry.*

IRAN, PAKISTAN AND INDIA

Most of the interior of Iran is a plateau surrounded by mountain ranges. The plateau, all of which is over 500 metres (1,650 feet) above sea level, is divided into two major areas: the salt desert, or Dasht-e Kavir, in the north and the Dasht-e Lut to the south. The Dasht-e Kavir is an inhospitable region, parts of which are completely uninhabitable. Salt plasters the surface, rendering cultivation impossible, and in places even travel is hazardous. Salt crusts cover areas of mud which conceal deep subterranean channels. The fragile structure of the surface is extremely dangerous.

The Dasht-e Lut is very different to the Dasht-e Kavir, being largely covered with loose sand and desert pavement (small stones). Both regions, however, suffer from extremes of climate. Summer temperatures reach over 50°C (122°F), while the winter temperature can drop below freezing. The temperature variation in the central Iranian plain is exacerbated by the high winds common to the area.

Average annual rainfall exceeds 600 millimetres (24 inches) only in the highlands of the west and the north. In the central plateau, rain is much more scarce, with an annual average of 200 millimetres (8 inches) in the southeast and less than 100 millimetres (4 inches) elsewhere.

Tehran

Tehran, the capital of Iran, encapsulates the environmental problems of urban settlement on the fringe of the desert. The city lacks an easily available water source and has long since exceeded its supposed maximum population of 5.5 million, identified by a development plan in the 1970s. An influx of people from the countryside, something that has occurred elsewhere in the Middle East, is just one of the factors that has caused the city's population to rise to over 8 million. Over-population in urban areas of Iran, and in Tehran in particular, has led to a reduction in the available clean water supplies in many areas of the country. Political events have slowed attempts to find solutions to these problems.

Below *A satellite image showing the Chagai Hills (running from lower left to upper right of the picture) on the border between Afghanistan and Pakistan. The dune field in the upper left-hand corner is a part of the Dasht-e Margo, which lies in southern Afghanistan. The dunes that are visible in the lower right-hand corner of the picture form part of a small sandy desert that is sometimes called the Dasht-e Tahlab. These deserts lie between the deserts of Iran and the Thar. Rivers from the Chagai Hills flow both northwards and southwards, draining finally into saline marshes and semi-permanent lakes.*

The Great Indian Desert

Pakistan and India are separated by a vast natural barrier, the Thar, or Great Indian, Desert. The Thar covers an area of some 200,000 square kilometres (77,000 square miles) and is well known for its sand dunes, some of which can reach a height of 150 metres (500 feet).

Average annual rainfall in the region ranges from about 100 millimetres (4 inches) in the west, to 500 millimetres (20 inches) in the east. The lack of rain combined with strong winds of up to 150 kilometres per hour (90 miles per hour), particularly in May and June, not only creates frightening sand storms, but also leads to fertile soil often being covered over with sand. Unlike many deserts, the Thar also lacks a good groundwater resource. What little groundwater there is lies very deep and is usually too saline to be useful.

Because of these harsh factors and summer temperatures of 50°C (122°F) very few people formerly lived in the region. Those that did raised sheep and cattle in areas where there was sufficient water for grass. However, with the completion of the huge Indira Gandhi Canal scheme in 1986, several irrigation projects have been started. These have attracted more pastoralists and farmers to the region.

▨	Desert
▨	Semi-arid
——	Main roads
——	Railways
——	Rivers
·-·-·	Seasonal rivers
◌	Seasonal lakes
≈	Salt flats
●	Oil and gas fields
——	Pipelines
◆	Mining/mineral exploitation
●	Large cities
○	Important desert towns/settlements
·—·—·	International boundaries
— —	Disputed boundaries

Below *The Indo-Iranian arid region stretches from the Caspian Sea southeast to the Rann of Kutch, and incorporates a number of distinct desert areas. The largest of these is the Thar Desert, which is separated from the others by the wide, fertile valleys of the Indus River and its many tributaries. Since ancient times, the Indus valley has been a major centre of human settlement.*

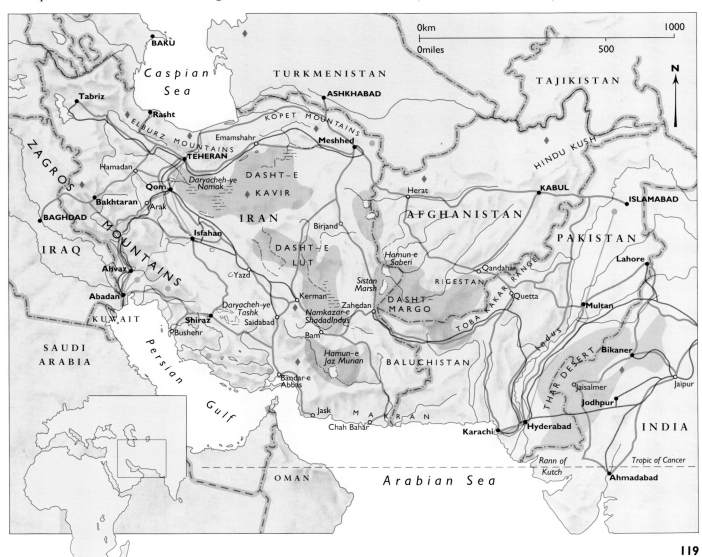

NATIONALITY AND RELIGION

The human and political landscape of the Middle East is a constantly changing, unstable mix of religions and national identities. This blend has contributed to numerous wars and movements of people, as well as a legacy of international disputes such as the Arab-Israeli conflict. Some of the most hotly disputed areas are also among the emptiest, but their oil fields have made them the focus of global strategic interest. A dispute between Iraq and Kuwait over the Burgan oil field was one factor that led to the Gulf War of 1991.

The Middle East saw the birth of three major monotheistic religions – Islam, Christianity and Judaism – that are still important today. Until the rise of Islam in the AD 600s, Christianity was the religion of much of the region. However, the Middle East has always been very divided, if only by the vast, inhospitable distances between peoples, and Christianity never exerted much influence over the desert nomads of Arabia. Islam itself is divided into two main branches, each subdivided into numerous sects.

Above *Tehran, 1980. Student activists demonstrating inside the captured United States Embassy hold aloft pictures of their spiritual and political leader, the Ayatollah Khomeini. The emergence of militant Shiite Islam under Khomeini's leadership provided a focus for strong anti-Western (especially anti-American) feelings that had developed in Iran during the previous half century.*

National identity

Nationality is a term more difficult to define than religion. It may denote ancient communities, such as the Egyptians and Persians, or it may indicate countries such as Libya and Jordan, which have only recently come into existence. Lebanon's mosaic of sects, communities and clans was given national status after the Second World War to further Western political goals in the region. Iraq's Kurdish minority represent a nation in terms of their common language, culture and homeland, whereas only religious identity and practice distinguish the Shiite Muslims in the marshes of southeast Iraq from their fellow Iraqis.

The Middle East is, above all, the home of the Arab people. The Arabs are united by their common language and heritage, but are divided into a northern, Mediterranean group who claim descent from Adnan, and the South Arabian group found in Yemen, the Hadhramaut and Oman. This distinction was perpetuated in the days of the Arab empire, leading to civil strife and dynastic squabbles. The decline of the Ottoman empire in the early 1900s gave birth to Arab nationalism. By the end of World War II, most of the Arab states had won their independence. During the 1960s, the Egyptian president, Gamal Abdel Nasser, led the unsuccessful drive for Arab political unity known as Pan-Arabism. Some Muslims dream of an age which will bring about the demise of nationality through the achievement of a grand pan-Islamic vision. Many Arabs still dream of creating some form of pan-Arab state.

Modern Israel is a relatively very recent state, coming into existence officially in 1948, yet it enshrines Jewish dreams of a return to the homeland that date back two thousand years. Within Israel, there are Arab Muslim, Christian and Druze minorities who are Israeli by nationality, but whose loyalties may lie elsewhere. As a result of Israel's occupation of the West Bank of the Jordan River (including eastern Jerusalem) and the Gaza Strip during the Six-Day War of 1967, Israel rules a rebellious Palestinian Arab population, who also cherish dreams of nationhood.

Top *American troops drive across an apocalyptic landscape in the aftermath of Iraq's 1990 invasion of Kuwait. The desert sands are stained black with spilled oil. The flames on the skyline are oil wells set on fire by the Iraqi army retreating in the face of the counter-invasion by forces of the Arab-American-European coalition.*

Above *Women warriors march in the streets of Tehran, Iran. The flowers in their rifles do nothing to soften this sinister image. Despite being accorded equal status with men as soldiers of the revolution, all Iranian women are required by strict religious laws to cover their heads and bodies with the traditional long, hooded cloak.*

Militant Islam

Islam is divided into two branches: Sunni and Shiah. Sunni Muslims represent 80 per cent of the Islamic world, but Shiites predominate in Iran. In 1979, the rise of Shiite militancy in Iran led to the overthrow of the monarch, the Shah, and the establishment of a religious government led by the *imam* (religious leader) the Ayatollah Khomeini. Iran's Islamic revolution sent shockwaves throughout the region. Iraq's Shiite minority are Arab, but their Shiite neighbours in Iran are Persian, which brought about a conflict of religious and national loyalties during the Iran-Iraq war (1980–88). In largely Sunni Afghanistan, ethnic divisions between the Pushtu-speakers and the Persian-speaking Tajiks have shattered the fabric of unity established during the struggle against the Soviet army.

Within the nomadic communities, charismatic religious movements have led to religious and political changes. The Almoravid movement, which began among the Berber nomads of the Western Sahara in the AD 1000s, resulted in the building of a North African empire. In the 1700s, the puritanical Wahhabi sect arose amongst the Bedouins of the Najd region of the Arabian Peninsula. In time, it brought about the establishment of the present Saudi ruling house.

CENTRAL ASIA

Central Asia is covered by the five Asian republics of the former USSR, and occupies an area two-fifths the size of Europe. The region is fringed by the Russian Federation to the north, the Caspian Sea to the west, the Kopet-Dag of Iran and Afghanistan to the south, and the Tien Shan, Pamir and Altay mountains to the south and east.

Most of the region consists of a depression containing the Aral and Caspian seas, and a series of plateaus and plains sloping gently to the Caspian. On its eastern shores, the Caspian breaks up into a number of salt lakes, much the most notable of which is Zaliv Kara Bogaz Gol. There are two other major areas of water in the region: the Aral Sea and Lake Balkhash. Otherwise the terrain is mainly desert and semi-arid sub-tropical terrain.

The two main deserts are the Kyzyl Kum (Red Sands) in Uzbekistan and the Kara Kum (Black Sands) of Turkmenia. There is also the Muyunkum Desert in south Kazakhstan and the Barsuki Desert north of the Aral Sea. The northern part of the Kyzyl Kum desert is a steppe area which has its annual spring abundance of vegetation scorched off by the summer heat. The southern part is a semi-desert plateau.

The two main rivers in the region are the Amu Darya and the Syr Darya, which rise in the mountains to the southeast of the region and flow generally northward to the Aral Sea. Extensive irrigation provided by these rivers has resulted in development of fertile land in the Ferghana valley of east Uzbekistan, southern Kazakhstan, and the northern fringes of Khirghizia descending to the Kazakhstan steppes. However, as rain is irregular, the Khorezm oasis of Khiva in Uzbekistan is being encroached by the desert, and many old watercourses have been dry for centuries.

Reservoirs have been created to secure expanding irrigation. A lake 500 metres (1,650 feet) long has been created at Bukhtarm on the upper stretches of the Irtysh in Kazakhstan, while the damming of the Amu Darya at Leninabad has formed another lake, the "Tajik Sea". Another notable artificial feature is the 1,350-kilometre (850-mile) Kara Kum Canal along the southern Turkmenian border.

The region lies in the largest continent at a relatively northern latitude, and there is no barrier between it and the arctic, resulting in cold winters and scorching summers. The snow of the mountain fringes feeds water sources with its run-off in the spring and summer. High winds combined with the desiccation of the Aral Sea, have caused serious salt storms in the region. Every year, over 40 million tonnes (tons) of salty grit is taken by the wind from the Aral Sea.

In the desert areas only the hardiest of vegetation can survive, for example, tamarisks (*Tamarix* spp.), the kok sagyg (which produces latex) and solianka, a bushy shrub with bright red flowers. The desiccation of the Aral Sea has resulted in increased salinity and this has caused the death of the 24 species of fish native to the sea. In addition, the boars (*Sus scrofa*), deer and egrets that the Aral Sea once supported have also disappeared. In the fertile areas, the main vegetation is commercial crops – predominantly cotton – with substantial numbers of fruit trees in the Fergana Valley and melon crops growing along the main river valleys.

Below *A satellite photograph (taken in 1985) looking southward from the top of the Aral Sea. When inverted, the image is not unlike most of the standard depictions of the Aral on maps (compare with the map opposite). The shoreline of this inland sea was first charted accurately during the 19th century. Until the 1960s, this depiction remained a reasonable representation of reality. By the early 1970s, however, shrinkage* *was becoming all too apparent. In the foreground, the island (depicted on maps) at the northern end of the Aral has become a peninsula that is extending to the opposite shore, which will create several small lakes. At the top right of the photograph, the separate body of water that can be seen was once a part of the Aral. The retreat of the water is most pronounced along the shallower western (right-hand in the photograph) side of the sea.*

Right *A satellite photograph of the Aral Sea (taken in 1992) showing the same scene as in the picture above, but taken from a higher altitude at a near-vertical angle. The Aral looks little like its conventional map image. More than 40 per cent of the open water area has been lost, and the shape has altered almost beyond recognition. Extensive areas of dry sea-bed are exposed and the old shorelines are obscured by drifts of alluvial and wind-blown silt that now cover much of the landscape. Given the current rate of shrinkage, any new maps that might be drawn of the Aral on the basis of its present* *shoreline will be significantly out of date in less than a decade. With a maximum depth (pre-1960) of less than 68 m (225 ft), the Aral was always a shallow sea, and had been shrinking very slowly since the end of the last Ice Age. Totally land-locked, and with a high rate of evaporation, salinity was already high before large-scale Soviet engineering brought about the present catastrophe. Ironically, the Aral Sea derives its name from an old Kirghiz phrase meaning "sea of islands"; these islands are now isolated bits of raised ground amid a sea of sand, silt and salt.*

RUSSIAN FEDERATION

Samara

Magnitogorsk

Petropavlovsk

Omsk

Saratov

Uralsk

Orsk

Kustanay

Tselinograd

Semipalatinsk

Aktyubinsk

Tobol

Ishim

Irtysh

Karaganda

Volgograd

Ural

NARYN
SANDS

Emba

Celkar

Turgai

Tengiz
Lake

Solonchak
Ghalkarteniz

Dzhezkazgan

Balkhash

Guryev

Astrakhan

BARSUKI
SANDS

Aralsk

KAZAKHSTAN

Lake Balkhash

SARY-
IZHIKOTRAU
SANDS

Lake
Alakol

Volga

Aral
Sea

BETPAK-DALA DESERT

Kzyl Orda

Chu

MUYUNKUM SANDS

TAUKUM
SANDS

Groznyy

▼-28m

Shevchenko

UST-URT
PLATEAU

KYZYL KUM SANDS

Syr Darya

Dzhambul

ALMA-ATA

Caspian

Nukus

Chimkent

KIRGHIZ RANGE

BISHKEK

Issyk-Kul
Lake

Sea

TASHKENT

Namangan

KIRGHIZIA

CHINA

BAKU

Amu Darya

UZBEKISTAN

Andizhan

Kokand

Krasnovodsk

KARA KUM SANDS

Bukhara

Samarkand

TURKESTAN RANGE

TIEN SHAN

TURKMENISTAN

Chardzhou

Karshi

TAJIKISTAN

▲7,134m

N

Kara Kum Canal

DUSHANBE

PAMIRS

ASHKHABAD

Mary

KOPET MOUNTAINS

Termez

IRAN

Meshhed

AFGHANISTAN

TEHERAN

0km — 1000

0miles — 500

	Desert
	Semi-arid
——	Main roads
——	Railways
——	Rivers
—·—·—	Seasonal rivers
◌	Seasonal lakes
◆	Mining/mineral exploitation
●	Oil and gas fields
——	Oil pipelines
●	Oasis
●	Large cities
○	Important desert towns/settlements
—·—·—	International boundaries

Above The deserts of Central Asia are cut off from the rest of the world by barren mountain ranges. The aridity of the area is underlined by the comparatively large number of named sandy deserts. The Aral Sea once marked the western boundary of the desert region, but unless the Aral can be restored to its former size, the Barsuki, Kara Kum, and Kyzyl Kum sands are likely to spread and fill the whole of the Aral Sea region.

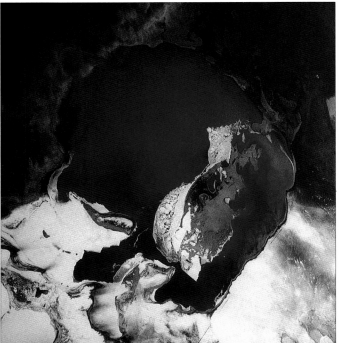

PEOPLE AND PROBLEMS

Central Asia is a predominantly Muslim area, with native populations consisting mainly of Tadjik (Indo-European) and Turco-Mongol stock. There is also a sprinkling of Muslim Uigur people who fled from repression in northwestern China. Additionally, substantial groups of ethnic Germans from the Volga region, as well as Trans-Caucasians, Ukrainians and Russians, resettled in Central Asia, particularly during the rule of Stalin, creating a very mixed population.

Industry and agriculture

Economic activity is a mixture of agriculture, the mining and processing of a variety of minerals, and energy-production industry. The area is rich in mineral resources, with coal in the Karaganda region of Kazakhstan, lead and zinc in the Altay Mountains, iron and copper around Lake Balkhash and Almalyk, and iron ore together with the associated production of steel in Tashkent.

The damming of the Syr Darya and Amu Darya, and the creation of several artificial lakes have led to considerable production of hydroelectric power in the mountains, although this is still an under-developed resource. Oil and natural gas reserves are found throughout the region, especially in the Caspian and Bukhara areas of Uzbekistan and in Turkmenia, and are also not exploited to their full potential.

The main crop is cotton – the "white gold" – which has made the region famous and drained its environment. The industry began in the nineteenth century when the American civil war cut exports from North America, creating a European demand for cotton from other sources. New irrigation projects began in 1918, and at the time the Soviet Union came to an end, the cotton industry was producing two-thirds of all Soviet cotton. As the industry developed, however, it relied increasingly on intensive irrigation and heavy subsidies from central funds to maintain production.

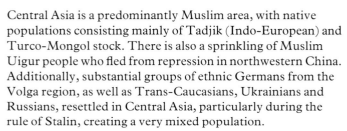

Above *Uzbek and Turkmen women gather around the Holy Well at Tash-Khaili Palace, Khiva, in western Uzbekistan. In ancient times, the Amu Darya River flowed into the Caspian Sea, and so provided the Khiva Oasis with much greater supplies of water than it receives today. Since large amounts of water began to be taken from the river, the Khiva region has become increasingly desiccated.*

Top left *A field worker picking cotton in Uzbekistan. As well as being a particularly thirsty crop, "white gold" also requires constant attention and intensive labour, especially at harvest time. Mechanical harvesting is possible, but the equipment is sophisticated and expensive. Under the centralized Soviet economy, such mechanization was considered to be an unnecessary investment.*

The desiccation of the Aral Sea

As a result of irrigation projects, the Aral Sea lost about 40 per cent of its surface area between 1960 and 1990, amounting to some 28,000 square kilometres (11,000 square miles). Diverting water for agriculture disturbed the balance between inflow to and evaporation from the sea. At current rates, the sea will shrink to two-thirds its 1990 size by the year 2000.

The desiccation of the Aral Sea poses a major environmental threat to the region and its population of 35 million people. The drop in the level of the sea has been accompanied by a corresponding drop in the quality of water. Furthermore, much land in the irrigated areas has been affected by salts rising to the surface in the naturally saline desert soil – a side effect of irrigation.

The northwestern region of Uzbekistan, which borders the Aral Sea, has suffered most directly from the environmental changes caused by the desiccation of the sea. There has been a significant increase in respiratory and eye diseases that are linked to the increasing amounts of salt and other sea-bed materials in the air. Childhood diseases associated with the increased aridity are also a problem.

The effects of desiccation have spread to many parts of the economy. The Karakalpak region once had a flourishing fishing and canning industry. As the sea receded so this industry failed and left the fishing centres landlocked.

Above *Fishing boats lost in a sea of sand near the former port of Muynak, Uzbekistan. Thirty years ago, Muynak was a thriving fishing port on the Aral Sea, and boats brought back bumper catches. Today, the sea has receded and the fish are long gone. Only these sand-blasted hulls remain as a strangely unreal reminder of a catastrophe that has blighted a whole region.*

It is the diversion of river water for irrigation purposes that has caused the present shrinking of the Aral Sea, but it is possible that this has accelerated a natural, if very much slower, process. The sea's level was considerably higher less than a million years ago. Records from historical times show that the Aral Sea had been shrinking, if slowly, until about 1880.

The problem of the Aral Sea is so great that drastic measures will have to be taken if it is not to dry completely. One solution suggested is to divert water from the Ob and the Irtysh rivers to feed the sea. Environmentalists are opposed to this, because the scheme could compound the region's problems. More realistic are the projects to strengthen irrigation canals and use water more effectively. In addition, attempts are being made to save delta lakes and turn some into fish farms to salvage the fishing industry.

On top of problems caused by irrigation, the cotton-growing areas suffer from the heavy use of pesticides, which has restricted much of the land to cotton growing (rather than food crops) for safety reasons. The use of Butifos, which defoliated cotton plants for ease of picking, and the continued use of DDT, despite bans, has poisoned the land.

THE GOBI DESERT

The Gobi Desert is the world's fifth largest desert. It takes its name from the Mongolian word meaning "waterless place". The Gobi stretches across southeast Mongolia and northern China, covering an area of approximately 1,295,000 square kilometres (500,000 square miles), and extending 1,610 kilometres (1,000 miles) from east to west and 910 kilometres (565 miles) from north to south. Rainfall in the Gobi averages 50–100 millimetres (2–4 inches) annually, although it is higher in the northeast of the region. Temperatures vary from −40°C (−40°F) in January to 45°C (113°F) in July.

The Gobi actually comprises several distinct arid areas. In the west lies the Taklimakan (or Ka Shun) Desert. It is bounded by the Tian Shan mountain range in the north and west, and the Kunlun and Atlun mountains to the south. The Taklimakan is a vast sea of dunes developed on an undulating, elevated plain, which rises to 1,524 metres (5,000 feet). The dunes are interspersed with a complex series of hills and scarps, which can rise 300 metres (980 feet) above the plain. Salty, dry lakes lie between the dunes, some of them eroded into fields of fantastic yardangs (wind-sculpted hillocks). Northeast of the Taklimakan lies the Junggar Basin, nestling between the eastern spurs of the Mongolian Altai and the eastern extremity of the Tian Shan Mountains. Its hills and low mountain ridges are dissected by ravines at its fringes. The Junggar Basin runs southeast into the Trans-Altai Gobi, bounded by the Mongolian Altai and the Gobi Altai in the north and east and the Atlun range in the south. The basin can be further subdivided into two distinct portions. In the east, it is a sharp rugged plain broken by a mountainous spur that extends 10 kilometres (6 miles) into the plain. The western portion is also a plain, across which meander dry river beds, and the central part of which becomes increasingly fragmented by mesas and dry gullies.

The majority of rivers in the Gobi flow only when rains fall – mostly in the summer. Rivers that flow into the region from the surrounding mountains quickly disappear into the dry ground. The eastern Mongolian Gobi region has more groundwater than the other parts of the desert. The water comes to the surface in small lakes and springs, but vegetation, even here, is still very sparse. Other distinct regions of the Gobi include the Alashan Desert, which is bordered by the Quilian Mountains to the southwest and the Huang He River to the east. To the east of the Huang He lies the Mu Us (or Ordos) Desert.

Life in the Gobi

The vegetation of the Gobi consists mainly of salt-resistant plants. Species such as succulent grass (*Enchinochloa* spp.), tamarisk (*Tamarix* spp.) and stunted willow (*Salix* spp.), are sparsely distributed. In the semi-arid regions, vegetation is slightly less sparse and does not need to be so hardy to survive; it includes such species as timuriya (*Timouria villosa*), and feather grass (*Stipa pennata*). Notable animals of the Gobi include camels (*Camelus bactrianus*), ass kulans (*Equus asinus ferus*), Przewalsky's wild horses (*Equus przewalskii*), gazelles (*Procapra* spp.) and ground squirrels.

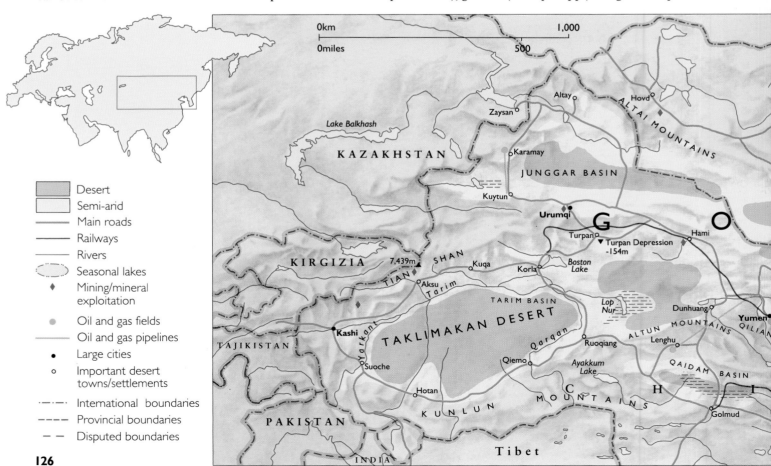

Key

- Desert
- Semi-arid
- Main roads
- Railways
- Rivers
- Seasonal lakes
- ♦ Mining/mineral exploitation
- ● Oil and gas fields
- Oil and gas pipelines
- ● Large cities
- ○ Important desert towns/settlements
- —·—· International boundaries
- —— — Provincial boundaries
- — — Disputed boundaries

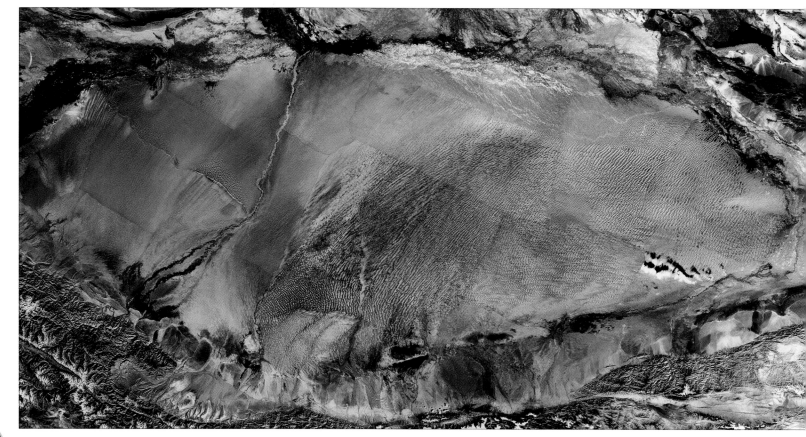

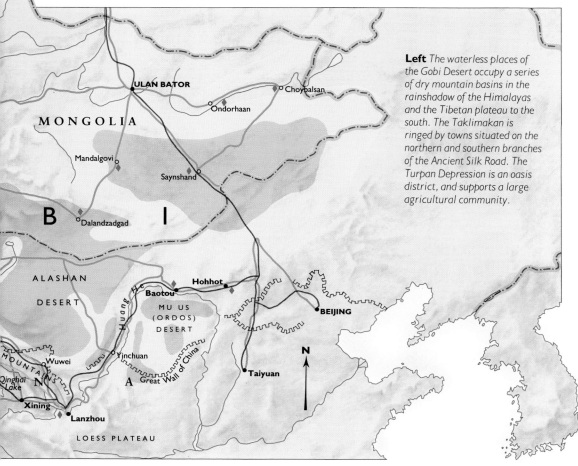

Left The waterless places of the Gobi Desert occupy a series of dry mountain basins in the rainshadow of the Himalayas and the Tibetan plateau to the south. The Taklimakan is ringed by towns situated on the northern and southern branches of the Ancient Silk Road. The Turpan Depression is an oasis district, and supports a large agricultural community.

ULAN BATOR

○ Choybalsan

○ Ondorhaan

MONGOLIA

Mandalgovi ○

Saynshand ○

B I

○ Dalandzadgad

ALASHAN

DESERT

Hohhot

Baotou

MU US (ORDOS) DESERT

● BEIJING

Huang He

N

Wuwei

Yinchuan

Great Wall of China

Taiyuan

MOUNTAINS

Qinghai Lake

Xining

Lanzhou

LOESS PLATEAU

PEOPLES OF THE GOBI

The Mongolian language has 33 different words to describe the Gobi. These words classify types of terrain in terms of the vegetation that is growing there, and reflect not only the subtle diversity of the terrain, but also the close affinity that the peoples of the Gobi have with their inhospitable environment. Like most arid regions of the world, the Gobi is very sparsely populated.

The people of the Gobi Desert are predominantly Khalkha Mongols, who make up about 80 per cent of the population of Mongolia. At the fringes of the Gobi, in the Inner Mongolian Autonomous Region of the People's Republic of China, Mongols are in a minority and ethnic Chinese compose the majority of the population.

The semi-nomadic pastoral lifestyle of the Khalkha Mongols has remained relatively unchanged for centuries. The main distinction between the Khalkha Mongols and other herdsmen in the region is that, because of the great harshness of their environment, they predominantly herd two-humped Bactrian camels. This species is better suited to the arid environment of the Gobi than is the horse, and is hardier than the one-humped Arabian camel (sometimes known as the dromedary). The 600,000 or so camels of the Gobi are not only essential for transporting the homes of the semi-nomads from pasture to pasture, but they are also sheared for their thick wool.

Although camels are numerous in the Gobi, sheep and goats are by far the most common livestock kept by the herdsmen. The sheep are particularly valuable as a source of meat, and the goats provide high-quality cashmere yarn. In the lusher semi-desert in the southwest of the region, herdsmen are also able to rear long-haired cattle.

Right *The snow-capped Tian Shan mountains tower above sparse grassland and dune fields in Xinjiang Province, China. The portable home (ger), the tethered pony and the scattered flock represent a nomadic way of life that has remained virtually unchanged for several thousand years.*

Inset right *Mongol elders squat outside a tightly wrapped ger. The domed Mongol tents are often mistakenly called yurts, but this term refers to an accompanying open enclosure. The tent has a wooden framework and is covered with layers of hides and woven fabric, which are tied in place.*

Below *A caravan of Bactrian camels makes its way across the almost featureless gravel plain around the Turpan depression. Although camels are still widely used in the Gobi Desert, motor vehicles are becoming increasingly common, as is indicated by the wheel tracks that bisect the caravan's route.*

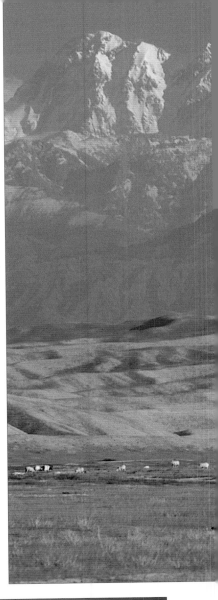

On average, nomadic herdsmen move about 10 times a year. To make travelling easier, the herdsmen and their families live in lightweight tents, known as *ger*, which are made out of felt. Nomadism takes place on all scales, from short moves for fresh pasture to longer seasonal migrations to escape the worst climatic extremes. The nomads of the Gobi usually live in family groups. The number of people in such groups is limited, and so this way of life helps to keep concentrations of people and livestock low in any particular area, thus not overstretching the desert's resources.

Where there is sufficient groundwater, as in the oases, some crops, such as melons and onions, are grown. However, the growing season is short and winter arrives early (in September) and is extremely harsh.

Natural resources
The only oil ever discovered in the Gobi was around the town of Saynshand in the northeast of the region. The reserves were developed by the Chinese, and explorations have been recommended at deeper levels. However, the fact that the area has little water to support large populations, as well as it being extremely isolated, could limit the economic viability of oil production. In some areas, salt and light metal ores are mined. Other products of the Gobi include scant quantities of coal, semi-precious stones and aromatic woods.

Historically there were a fair number of cultivated oases in the Gobi, which provided rest points on caravan trails. These settlements generally flourished until their populations became too great for their water supply. Today, the main barrier to development is still the lack of water.

Mongolia as a whole has only recently begun to emerge from a period of prolonged economic stagnation and decline. There are a few small towns on the communication routes that cross the Gobi, such as along the Beijing-Ulan Bator railway. In addition to this, there are regional administration centres, which provide basic services for the widely scattered herding populations. One such service is a boarding school system for the children of the herdsmen. The system allows the children to receive a good level of education without it being disrupted by the nomadic lifestyle of the parents.

In recent decades, the Chinese government has implemented a series of programmes in the southern parts of the desert to "reclaim the Gobi". Among these are projects experimenting with rice cultivation and viticulture. These are mostly at an experimental stage and are not economic in terms of returns on investment (vines are grown in greenhouses, for example) or in the use of the large amounts of scarce water needed. Other schemes include planting migrating sand dunes to prevent their continued advance on the fringes of the true steppeland of Inner Mongolia.

TIBET

Most of the high plateau of Tibet is desert. It is an autonomous region of China, located in the mountains of Asia. Covering an area of about 1,200,000 square kilometres (470,000 square miles), the region is more than twice the size of France, and most of it lies above 4,500 metres (15,000 feet). Along the whole length of Tibet's southern borders lie the Himalayas. To the northwest lies the Kunlun Mountain range, while to the northeast are the Tanggula Mountains.

The Tibetan plateau is a prime source of water for Central Asia, and groundwater and snow-melt feed the headwaters of the Indus, Brahmaputra, Salween, Mekong, Yangtze and Huang He river systems. However, much of Tibet is desert, apart from the slightly lower and wetter southeastern fringes, and receives less than 25 centimetres (10 inches) of rain or snow each year. The low precipitation is due mainly to the Himalayas, which act as a barrier to the monsoon winds, bringing torrential rain to areas only 640 kilometres (400 miles) south of Tibet. In addition, violent winds sweep over much of Tibet all year round, and these have a desiccating effect. The high altitudes ensure that the temperatures over most of Tibet are cold. Tibet's climate is characterized by very long winters with extremely low night-time temperatures, and short mild summers.

Surviving in the cold wilderness

Agriculture and pastoral nomadism are the two main activities. The waters of Tibet's major rivers are largely unavailable to the local population, who lack energy resources, manpower or technology to transfer it to their fields. Instead, small groundwater-fed streams are tapped through systems of narrow channels and ponds, and led into terraced fields which are mainly situated on relatively level valley floors. The traditional staple crop in most areas is barley, together with buckwheat, pulses and wheat, more recently potatoes, and in parts of the southeast, rice.

Between April and October, the animals (mainly cattle, yaks, sheep and goats) are grazed on upland pastures. Their dairy produce is an essential ingredient of the local diet. The animals retained over the winter are accommodated in the lower stories of the farmhouses and fed on hay, clover and other fodder. All the livestock, together with horses, mules and donkeys, are also used as beasts of burden.

Since the 1950s, most Tibetan-populated areas have been subject to attempts to integrate them economically, culturally and militarily with their neighbours. However, these attempts have been hampered by Tibet's remoteness and sparse population, as well as a cultural separateness based on its people's intense Buddhist faith. China has occupied Tibet since 1956, but has not succeeded in suppressing Tibetan cultural life. Projects for agricultural improvement have met with mixed success. A small local market and lack of surplus manpower make manufacturing uneconomic. A variety of minerals, including iron, manganese, magnesium, copper, lead and zinc, has been found in Tibet. However, information on the precise amounts of mineral resources available is scarce. Tourism has been greatly developed since the 1970s, but is liable to political interruption.

Above *Pilgrims climb a hillside near Gander monastery. In summer, the green valleys and distant mountains lend a tranquil air to Tibetan scenery. In the absence of dust storms, the daytime skies are permanently blue. In winter, low temperatures and icy winds can make venturing out of doors at all a fairly risky occupation.*

Top *Photograph taken from an orbiting spacecraft showing an oblique view (looking eastward) along the Himalayas. The barren Tibetan Plateau (to the left of the snow-capped peaks) contrasts strongly with the greener, and obviously more fertile and better-watered, landscape in Nepal and northern India (right of picture). Most of the moisture carried this far by the wind is deposited on the southern slopes of the Himalayas, with little carried over the mountains into Tibet.*

High mountain
desert

Main roads

Railways

Rivers

Lakes

● Large cities

○ Important desert
towns/settlements

—·—·— International boundaries

———— Provincial boundaries

— — Disputed boundaries

Below *The high mountain desert of
the Tibetan Plateau spreads out in
the rainshadow of the Himalayas,
which (like the plateau) were
elevated in the geological past when
India collided with the continent of
Asia. To the south, on the windward
margins of the Himalayas, rainfall
can be torrential. To the north lie
the deserts of Central Asia.*

THE GREAT PLAINS

Westward expansion by United States and Canadian farmers during the mid-1800s halted along a frontier that stretched almost due north from the southern tip of Texas to central North Dakota, and arched through southern Saskatchewan and Alberta to the Rocky Mountains. This line marks the western edge of the prairies, beyond which the lack of water and wood inhibited agriculture and settlement. These dry western plains, or "Great Plains", are sometimes described as the "Great American Desert".

The Great Plains comprise a region delineated by geology and landforms. They form a series of gently rolling surfaces which rise towards the west in low steps to an elevation of about 1,800 metres (6,000 feet). The region is semi-arid throughout, but there is marked gradation between the north, which has severe, snowy winters and mild summers, and the south, with its occasional snow and long, scorching summers. The rivers that rise in the snow-clad Rockies (the Saskatchewan, Missouri, Yellowstone, Platte, Arkansas, Canadian, and Pecos) all diminish as they flow eastwards, but they nourish adjacent groundwater supplies.

The Sonora and Chihuahua deserts

The dry landscapes that dominate the entire southwestern quarter of North America are perhaps best conceived of as three fairly distinct desert cores. Each is surrounded by shrub and grass-covered transitions, and is interrupted throughout by mountainous enclaves of woodland and forest.

The Sonora Desert nearly encircles the Gulf of California. It extends from sea-level northwards into the lower valleys of California and Arizona, eastwards into the Sierra foothills of Sonora, and in many places westwards across the peninsula of Baja California to the shores of the Pacific. The Chihuahua Desert lies at the relatively high elevation of 1,000–1,500 metres (3,300–5,000 feet), and occupies the open northern end of Mexico's Mesa del Norte. It is flanked to the east and west by zones of severely degraded grassland and woodland which rise up to the Sierra and to the temperate Valley of Mexico on the Central Plateau.

Both Mexican deserts, in contrast to the third desert region, in the Great Basin, are essentially subtropical and are characterized by exceedingly rich floras containing annuals, woody perennials, and succulents, including many species of cactus. True barrenness is confined to dry lake beds (*playas*, salt flats and clay pans), rare tracts of mobile sand, "badlands" formed on deposits of shale and mudstone, and localized scree slopes and outcrops of rock.

Mining water

In the Texas Panhandle, near Lubbock, more than 5 billion cubic metres (176 billion cubic feet) of water are used annually to irrigate fields of cotton and other crops. The water comes from some 50,000 wells sunk into the famous Ogallala aquifer, which is being recharged at only a fraction of the rate it is being used. The increasing cost of pumping from greater and greater depths has forced changes in crop patterns and farming practices, and there is still little prospect for sustained agricultural production.

Below A satellite image of the White Sands National Monument in New Mexico, alongside the Sacramento Mountains (right of picture). The pale clays and associated salt deposits are typical of the intermontane basins in the arid Southwest. The White Sands missile range, which lies west of Alamogordo, was the site of the world's first atomic explosion.

Right The Great Plains extend in a 500–600 km (300–360 mile) wide strip from the Mackenzie lowlands in northern Canada, to the Rio Grande in the Texas Big Bend. To the south, the landscape degrades into the deserts of Sonora and Chihuahua. The Great Plains have a number of notable surface features, including the badlands of South Dakota and the sand hills of Nebraska, which are the largest dunefields in the Americas.

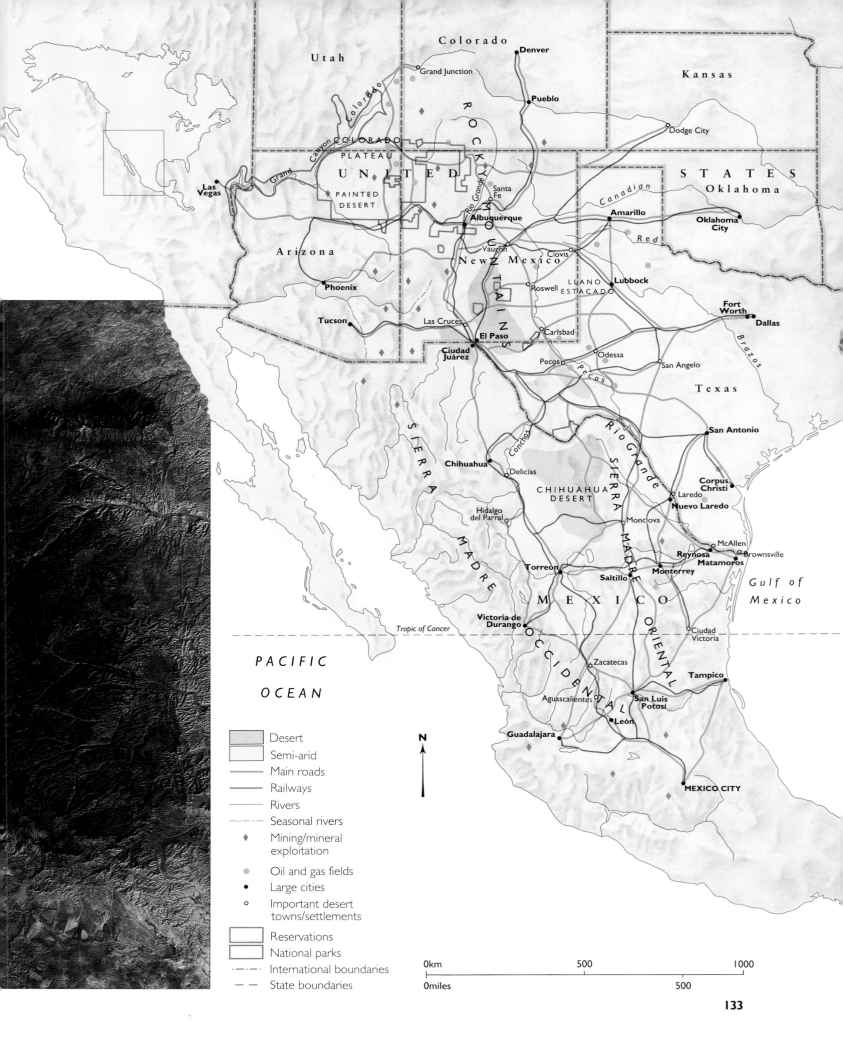

Utah

Colorado

Denver

Grand Junction

Pueblo

Kansas

R O C K Y

Dodge City

S T A T E S

Oklahoma

COLORADO
PLATEAU

UNITED

Las
Vegas

PAINTED
DESERT

Santa
Fe

Canyon COLORADO

Grand

Rio Grande

Albuquerque

Canadian

Amarillo

Red

**Oklahoma
City**

Arizona

Vaughn

New Z Mexico

Clovis

Lubbock

**Fort
Worth**

Phoenix

Roswell

LLANO
ESTACADO

Dallas

Tucson

Las Cruces

El Paso

Carlsbad

M O U N T A I N S

Brazos

**Ciudad
Juárez**

Pecos

Odessa

San Angelo

Pecos

Texas

SIERRA

San Antonio

Chihuahua

Conchos

**Corpus
Christi**

MADRE

Delicias

CHIHUAHUA
DESERT

Laredo

Nuevo Laredo

Rio Grande

Monclova

Hidalgo
del Parral

McAllen

SIERRA

Brownsville

Reynosa

Matamoros

Torreón

Saltillo

Monterrey

MADRE

*Gulf of
Mexico*

PACIFIC

MEXICO

OCEAN

**Victoria de
Durango**

Tropic of Cancer

OCCIDENTAL

ORIENTAL

Ciudad
Victoria

Tampico

Zacatecas

Aguascalientes

**San Luis
Potosí**

León

Guadalajara

N

MEXICO CITY

	Desert
	Semi-arid
	Main roads
	Railways
	Rivers
	Seasonal rivers
♦	Mining/mineral exploitation
●	Oil and gas fields
●	Large cities
○	Important desert towns/settlements
	Reservations
	National parks
	International boundaries
	State boundaries

0km   500   1000

0miles   500

133

GRASSLANDS AND HACIENDAS

The ubiquitous grass cover of the North American Plains fell under the plough more than a century ago. Tribes of indigenous peoples, particularly the Sioux, offered fierce resistance, but they were overwhelmed in the decade following the American Civil War (1861–65) by a nation left with immense military capacity and a surplus of adventurers and potential migrants. Crops and farming practices successful in the Midwest were initially applied to the Plains, but agricultural economics and "newfangled" machinery favoured cereals, particularly wheat, for export eastward.

Irrigated cropland is constrained by the region's limited water supplies and is confined to particular areas: narrow valleys of major rivers, especially the Yellowstone, Platte and Arkansas; the foothills of the Colorado Rockies, where some headwaters drain to the Plains; and various areas, from West Texas to South Dakota, underlain by the Ogallala aquifer.

Livestock ranching dominates areas of rough terrain and droughty soils – notably the "broken lands" of the western Dakotas and eastern Montana, and the Nebraska Sand Hills. It also prevails on the western and southern margins of the Plains, where precipitation is low and droughts are frequent.

The discovery of petroleum at the Spindletop gusher near Beaumont in 1901 fuelled a boom in the Permian Basin of western Texas. Sub-bituminous coal is mined from a dozen vast, open-pit mines in Wyoming's Powder River Basin.

Right *The wheeled arm of a centre-pivot irrigation system delivers precious water from an underground aquifer to a wheat field on the Plains. The gathering clouds overhead threaten a storm that could easily hammer flat the growing wheat with hailstones. Arable farming in this region is caught between the two almost opposing perils of too little water and the effects of sudden deluges of excessive amounts.*

Below *Branding time. In some ways, cattle ranching has changed little since the days when "the six-gun ruled the West", and modern technology has had relatively limited impact. Although modern-day cowboys communicate by radio, and are as likely to ride a jeep as a horse, it still takes up to five grown men to subdue one uncooperative half-grown animal.*

Below *The unearthly landscape of the White Sands National Monument, New Mexico. Trapped in a mountain basin, pale-coloured alluvial deposits have provided sands for an extensive dune field Lying between the dunes are exposed patches of desert floor with scattered clumps of grass. The shifting surface of the dunes remains free of vegetation.*

Northern Mexico

Spanish exploitation of Mexico's dry north dates from the mid-1500s, scarcely two decades after the conquest of the Aztecs. The quest for land, silver and souls to convert led quickly to the annihilation, assimilation or displacement of nomadic bands of indigenous peoples who inhabited the area. The semi-arid zones between the Sierras and the desert core contained the region's prized resources – minerals and livestock range – and provided the corridors for expansion. A few isolated outposts were founded far northward remarkably early, such as Santa Fe in 1609.

By the end of the 1600s, the frontier of settlement lay roughly along the current United States-Mexico border, and the essential characteristics of the rural economy and landscape were well established. There were scattered mining centres and small agricultural settlements, but the grasslands were dominated by huge, unfenced and virtually self-sufficient haciendas. These produced livestock – mostly cattle – for local consumption and for export to southern markets. The core deserts and rugged Sierras were generally avoided, but provided a refuge for remnants of indigenous peoples, such as the Sari in coastal Sonora, and the Tarahumara in the mountains of western Chihuahua.

With some notable exceptions, this pattern still persists. The United States is now the major market for Chihuahuan cattle, while copper production has outstripped that of silver.

The revolution of 1910 brought a degree of land reform, which dismantled many of the large haciendas. The Federal Government developed three important irrigated oases around the periphery of the Chihuahuan Desert – at points on the Rio Nazas and Rio Conchos, and along the lower Rio Bravo del Norte (the "Rio Grande" of Texas). Most of the winter lettuce and tomatoes consumed in the United States now come from irrigated fields on the coast of southern Sonora and northern Sinaloa, and Monterrey has become an important centre for steel production and a number of heavy manufacturing industries.

The most striking changes have occurred along the United States-Mexico border. Tourism, based initially on prostitution and the free availability of alcohol at a time when there were restrictive United States liquor laws, accelerated rapidly during World War II at communities adjacent to United States military bases. A more broadly based tourist industry has since thrived at most significant crossing points. Since initiation in 1965, the Border Industries Program, which takes advantage of United States' tax laws and low Mexican wages, has attracted more than 1,000 "maquiladora" factories, or assembly plants (largely American-owned), and 250,000 workers (entirely Mexican) to rapidly growing border communities. There are now a dozen major "twin" United States-Mexico settlements clustered along the international boundary.

THE GREAT BASIN

Most of the North American dry lands lie between the Rocky Mountains and a continuous wall formed by the Sierra Nevada of California and the Cascades Range of Oregon and Washington. The region is a good example of a mid-latitude rain-shadow desert, though the terrain subdivides the area.

Mountain and basin

The Rockies contain several sizeable dry regions, of which the Wyoming Basin is a good example. It effectively cuts the Rockies in half, and its gently undulating surface is covered by sparse grasses and sage (*Artemisia*). The San Luis Valley of southern Colorado is a high, dry basin roughly 50 kilometres (30 miles) across, surrounded by the peaks of the San Juan and Sangre de Cristo ranges. The valley floor is covered by brushy sage and greasewood (*Sarcobatus*), and contains irrigated oases and an area of salt flats.

The Great Basin is not a single depression, but a collection of basins, and is the classic example of "basin and range" topography. Death Valley National Monument includes spectacular examples of the terrain and its associated ecological conditions. Most of the basins are flat, and occasionally covered by shallow lakes. These soon evaporate, depositing clay, and precipitating salts. During the last ice age, these so-called *playas* formed extensive chains of lakes and inland seas. Utah's Great Salt Lake is a tiny remnant of Lake Bonneville, the dry, salt-encrusted bed of which now forms the Great Salt Lake Desert.

The ranges in Nevada are aligned north–south and are typically 100 kilometres (60 miles) or more in length. The upper slopes are frequently covered with pine and juniper stands. The intermediate zone between mountain base and basin floor is known as the "piedmont", and is made up of eroded material carried by streams. Piedmont zones comprise about 70 per cent of the Great Basin and contain its cover of shrubs, dominated in most places by greasewood, sage (*Artemisia* spp.) and saltbush (*Atriplex* spp.).

The plateaus

The Colorado Plateau is a wedge-shaped area of about 325,000 square kilometres (125,000 square miles) between the southern parts of the Rockies and the Great Basin, extending south into Arizona and New Mexico. Its geology and landforms are strikingly different from neighbouring areas. The region is underlain throughout by massive, ancient beds of marine sandstones, shales and limestones that have been deformed, uplifted, and deeply incised by the Colorado River and its tributaries. This is a land of extensive plateaus and flat-topped mesas (steep-sided tablelands). Colourful orange and red rocks are exposed in abundant cliffs and spectacular canyons, including the Grand Canyon itself.

The Columbia Plateau is a dry tableland covering some 130,000 square kilometres (50,000 square miles) of eastern Oregon and Washington, and the Snake River Plain of southern Idaho. It is covered with layers of solidified lava, which is 2,000 metres (6,600 feet) thick in places. The Columbia River and its major tributaries have carved canyons through parts of the plateau, forming ideal dam sites.

Below *A satellite image showing Las Vegas, Nevada, located in the indefinite borderlands between the Great Basin and the Mojave Desert. The city can be seen in the centre of the picture, surrounded by plumes of wind-blown dust. From the bottom right of the picture, the Colorado River runs north along the Arizona-Nevada border into Lake Mead, a large reservoir east of Las Vegas.*

Right *The Great Basin arid region covers most of Nevada and Utah, and extends northward and westward to include several smaller basins in adjoining states. Although its name suggests a straightforward drainage pattern, the situation is complex and subject to local variation. The Great Salt Lake (like many others) is saline, while nearby Utah Lake has fresh water.*

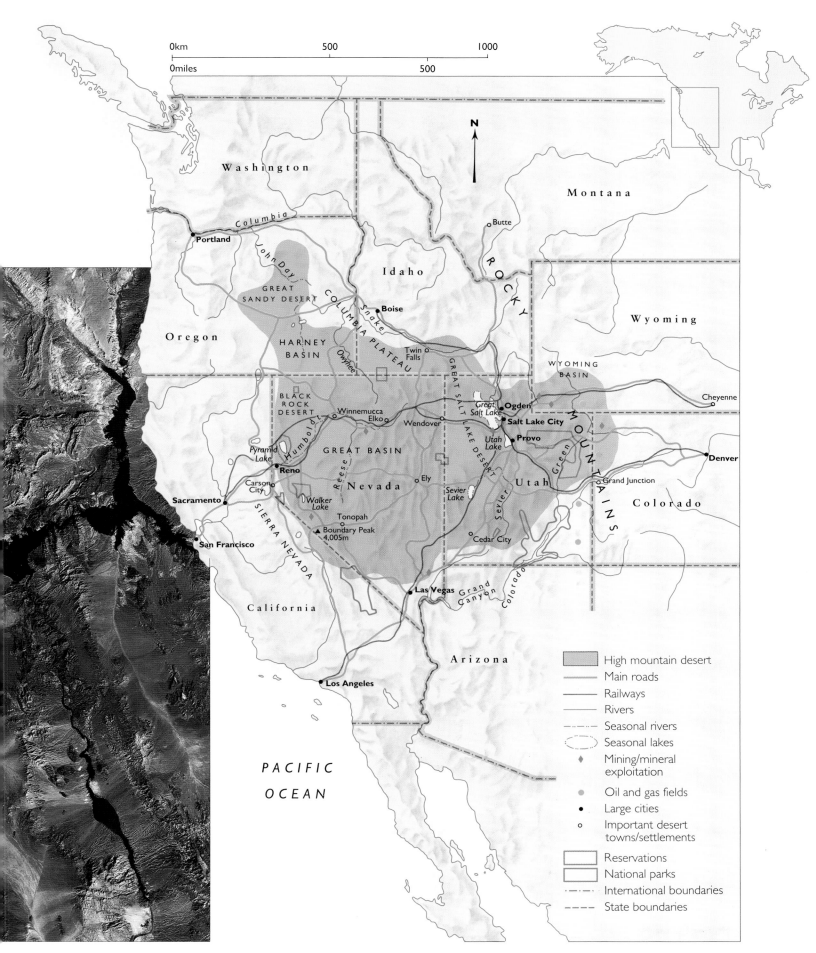

0km   500   1000

0miles   500

N

Washington

Montana

Columbia

○ Butte

Portland

John Day

Idaho

Wyoming

GREAT
SANDY DESERT

ROCKY

● Boise

COLUMBIA PLATEAU

Snake

Oregon

HARNEY
BASIN

WYOMING
BASIN

Twin
Falls ○

GREAT SALT LAKE DESERT

Cheyenne ○

Owyhee

BLACK
ROCK
DESERT

Great
Salt Lake

Ogden ●

Winnemucca ○

Elko ○

Wendover ○

Salt Lake City ●

Humboldt

Pyramid
Lake

GREAT BASIN

Utah
Lake

Provo ●

Green

MOUNTAINS

Reese

Reno ●

Carson
City ○

Ely ○

NEVADA

Sevier
Lake

Utah

Grand Junction ○

Denver ●

Sacramento ●

Walker
Lake

Sevier

Colorado

SIERRA NEVADA

Tonopah ○

▲ Boundary Peak
4,005m

Cedar City ○

San Francisco ●

California

Las Vegas ●

Grand
Canyon

Colorado

Arizona

Los Angeles ●

	High mountain desert
	Main roads
	Railways
	Rivers
	Seasonal rivers
	Seasonal lakes
◆	Mining/mineral exploitation
●	Oil and gas fields
●	Large cities
○	Important desert towns/settlements
	Reservations
	National parks
	International boundaries
	State boundaries

PACIFIC
OCEAN

A MODERN WILDERNESS

Right *Monument Valley, Arizona, is one of the most famous desert landscapes in the world, thanks to its appearance in numerous "Western" films. Although the towering buttes and rock pillars are especially spectacular, the barren rocky landscape is typical of that crossed by the early miners and settlers of the mid 1800s.*

The first European settlers to arrive in the intermontane deserts were miners and Mormon colonists. These two contrasting groups soon displaced the indigenous, nomadic population, which included tribes of the Navajo, Shoshone and Ute. Huge areas of land in the area are now owned and controlled by the United States federal government.

Following the California Gold Rush of 1849–59, a wave of prospectors moved eastward, combing the Great Basin and Rocky Mountains for signs of gold and silver. Numerous strikes gave rise to flourishing mining towns, but most of these communities founded on ore deposits proved to be short-lived. In Nevada, still known as the Silver State, the scars and debris of past mining activities are widespread. A few sizeable "ghost towns", such as Virginia City, have escaped fire and vandalism to become significant tourist attractions. Mining and processing of gold, silver, copper and other metallic ores are important in isolated localities today, but their regional economic significance is dwarfed by gambling, tourism and manufacturing.

Mormon colonies

The Mormons – members of the Church of Jesus Christ of the Latter-day Saints – arrived at the site of Salt Lake City in 1847. After the first years of hardship in the unfamiliar desert environment, the sect rapidly evolved social institutions, agricultural practices and colonial policies which proved remarkably effective for the expansion of their empire throughout the intermontane West. Wherever supplies of water could be developed, small groups of families were dispatched to establish villages. Green fields, orchards and pastures were sustained by irrigation, and surrounding deserts and woodlands served as common rangelands.

Some of the villages have been urbanized and lost their original character, notably Salt Lake City and Las Vegas. Many experienced growth and change as they acquired additional functions in government, tourism, forestry, manufacturing and mining. Scores were abandoned, especially in parts of southern Utah, as rangelands deteriorated and irrigation systems were eroded.

The public domain

Federally owned land – the so-called Public Domain – is especially prominent in the dry intermontane West, where it comprises 86 per cent of the state of Nevada and 66 per cent of Utah. Federal stewardship has profoundly influenced the use and conservation of these areas, with initially disastrous consequences. There were only minimal restrictions on use and there was a popular notion that any individual had rights to publicly held resources. This encouraged the removal of forests, indiscriminate slaughter of wildlife, and stocking of ranges far beyond their sustainable capacities. As a result of such exploitation, many areas had reached the peak of economic and ecological productivity before the end of the 1800s and have declined ever since.

Changes in public attitude and government policy have come slowly, but are now incorporated in hundreds of laws that carve the Public Domain into various types of management units with unique objectives for utilization, preservation and reclamation.

Inset top *A mobile seismic survey unit in Utah. Long after the original gold and silver prospectors had departed, the search for other forms of mineral wealth returned to the intermontane area. Since 1900, several deposits of oil, coal and gas have been found beneath the desert floor, and more are expected to be discovered in the near future.*

Above *A Navajo woman from Utah. The Navajo are native inhabitants of this region and are descended from peoples who built the Pueblo cities more than 1,000 years ago. Because of their settled lifestyle, the Navajo were relatively little affected by White settlers even up until World War II. Since then, traditions have declined.*

ARIZONA AND CALIFORNIA

The deserts of southeastern California are generally divided into two regions. The southern part, which lies at lower altitude, is really an extension of the Sonora Desert and is often called the Colorado Desert or simply the "Low Desert". The Sonora Desert covers much of southwest Arizona and northwest Mexico. The higher altitudes to the north of the Sonora, dominated by extensive stands of creosote bush (*Larrea tridentata*), are known to Californians as the Mojave Desert or "High Desert".

The Colorado River
Today, hardly a drop of water passes through the lower reaches of the Colorado River unless a valve is opened. The river that carved the Grand Canyon once had the power – during an extended flood 90 years ago – to escape through a small irrigation canal and create California's Salton Sea. But today, its waters are used, reused, and exported hundreds of miles to the major urban centres of Arizona and California. The keystone of confinement was the Hoover Dam (completed in 1936), which created Lake Mead. This was followed shortly by the Parker Dam (1938) with its reservoir, Lake Havasu, which is the source of the great Colorado River Aqueduct and the more recent Granite Reef Aqueduct conveying water to central Arizona. A series of smaller dams was constructed downstream to irrigate lands along the lower Colorado and to divert water to oases in the Imperial Valley and along the lower Gila River. The Davis Dam (1954) eased regulation and increased storage capacity. The last dam, Glen Canyon (1966), was built to slow siltation in Lake Mead.

The Colorado system controls floods, generates sizeable amounts of electricity, provides irrigation and municipal supplies of water and supports an important recreation industry. Ecologically, however, control of the river has been a disaster, altering river bank habitats along the lower valley and through the Grand Canyon. The proposed Bridge Canyon Dam would have backed water into the middle reaches of the Grand Canyon, but economic considerations and environmental groups halted the project.

Vegetation change
Since the mid-1800s, there have been significant vegetation changes in the Sonora Desert of southern Arizona. In some places, habitats have been obliterated, exotic species have been introduced, and the mix and composition of native species has been altered. Efforts to preserve vanishing river bank zones – especially groves of cottonwood (*Populus fremontii*) and willow (*Salix* sp.), and wet marshes – have been a major focus for local conservation groups.

The deterioration of the grasslands became apparent in the 1880s, soon after the large-scale introduction of cattle. Grass cover was greatly reduced and more than 20 plant species were introduced. Perhaps one-half of the 3,600,000 hectares (9,000,000 acres) of "mesquite land" in Arizona has appeared since about 1860. However, these changes are complex and are not necessarily negative. Many ranges throughout the western United States are today in far better condition than during peak periods of heavy grazing, which typically occurred before 1900.

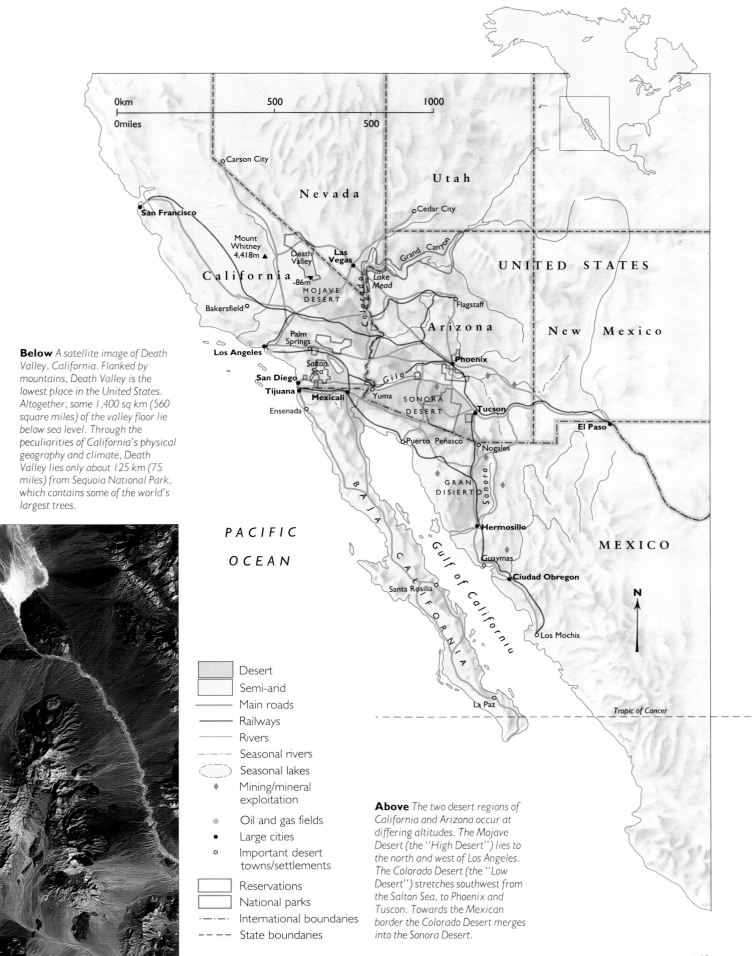

Below *A satellite image of Death Valley, California. Flanked by mountains, Death Valley is the lowest place in the United States. Altogether, some 1,400 sq km (560 square miles) of the valley floor lie below sea level. Through the peculiarities of California's physical geography and climate, Death Valley lies only about 125 km (75 miles) from Sequoia National Park, which contains some of the world's largest trees.*

Above *The two desert regions of California and Arizona occur at differing altitudes. The Mojave Desert (the "High Desert") lies to the north and west of Los Angeles. The Colorado Desert (the "Low Desert") stretches southwest from the Salton Sea, to Phoenix and Tuscon. Towards the Mexican border the Colorado Desert merges into the Sonora Desert.*

Legend:
- Desert
- Semi-arid
- Main roads
- Railways
- Rivers
- Seasonal rivers
- Seasonal lakes
- Mining/mineral exploitation
- Oil and gas fields
- Large cities
- Important desert towns/settlements
- Reservations
- National parks
- International boundaries
- State boundaries

Map labels:

United States, Mexico, California, Nevada, Utah, Arizona, New Mexico

Carson City, Cedar City, San Francisco, Mount Whitney 4,418m, Death Valley, -86m, Las Vegas, Lake Mead, Grand Canyon, Flagstaff, Mojave Desert, Bakersfield, Colorado, Palm Springs, Los Angeles, Salton Sea, Gila, Phoenix, San Diego, Tijuana, Mexicali, Yuma, Sonora Desert, Tucson, Ensenada, El Paso, Nogales, Puerto Peñasco, Gran Desierto, Sonora, Baja California, Pacific Ocean, Hermosillo, Guaymas, Ciudad Obregon, Santa Rosilia, Gulf of California, Los Mochis, La Paz, Tropic of Cancer

0km 500 1000
0miles 500

N

THE SUN BELT

Left *Phoenix, the state capital of Arizona, USA, a modern version of the traditional oasis city, where trees, lawns and swimming pools all thumb their noses at the arid desert climate. The water that supplies Phoenix comes from the Roosevelt Dam (completed 1911) on the Salt River, about 100 km (60 miles) to the northeast of the city.*

The state of California uses an annual total of 58 billion cubic metres (47 million acre feet) of fresh water. Around 40 per cent of this is pumped from groundwater reserves, principally in the Central Valley. The surface supply comes mainly from the western slopes of the Sierra Nevada where there are numerous reservoirs supplying a maze of canals that lead to croplands in the Central Valley. A smaller amount derives from the dry eastern slopes of the Sierras, mainly to the benefit of southern California. In most years, the Colorado River has contributed a far greater share to the total supply than California's legal allotment of 5 billion cubic metres (4 million acre feet) would suggest. As Arizona begins to utilize its share more fully, California's use will decline.

Huge amounts of water are transferred throughout the state, generally from the wet northern region to the dry southern region. The most spectacular transfers involve three great aqueducts that converge on the Los Angeles Basin: the Colorado River Aqueduct, the Los Angeles Aqueduct from the eastern slopes of the Sierra Nevada, and the California Aqueduct from the "Delta" region of the Central Valley near Stockton. The aqueduct from the Sierra Nevada to San Francisco is also impressive, as is the All American Canal that feeds Colorado River water to the Imperial Valley.

About 82 per cent of the total supply, including both groundwater and surface water, is used to irrigate crops. Roughly 85 per cent of the state's 40,000 square kilometres (15,000 square miles) of irrigated cropland is in the Central Valley, but there are other important areas: the Imperial Valley and Palo Verde Valley along the lower Colorado River, both of which lie in the Low Desert; and the Salinas Valley and Oxnard Plain, which lie between Los Angeles and San Francisco in moderately humid coastal locations. Several valleys near Los Angeles continued to produce irrigated crops until they were displaced by urban growth during the 1960s.

Supplies of groundwater were once abundant in the deep sediments of the Los Angeles Basin and other coastal valleys. However, these have been seriously depleted by a century of crop production and urban growth to the point where salt water is now spreading into aquifers along the coastal margins of the basin. To halt the intrusion of salt water, a chain of injection wells has been established, which create a barrier of fresh water. During periods when reservoirs along the Colorado are full, "excess" water is imported to the coastal margins in an attempt to recharge aquifers.

The San Joaquin Valley

Much of the southern half of California's Central Valley was a desert before irrigation. Water from the wet western slopes of the Sierra Nevada and from wells drilled deep beneath the valley floor have helped to turn the San Joaquin Valley into a major contributor to California's fantastic agricultural output. However, rates of groundwater consumption are now spectacularly excessive. This has produced rapidly falling water tables and huge areas where the land has subsided by as much as several metres. The supply of surface water is also a problem because California has lengthy dry periods.

The Navajo Reservation

More than 90,000 Native Americans live on Arizona's share of the Navajo Reservation, which also includes adjacent strips of New Mexico, Colorado and Utah. This "Four Corners" section of the Colorado Plateau is well known for its scenery and archaeological interest. But the area is also characterized by contemporary features of Navajo culture. Herds of sheep and goats graze on tufts of grass amongst the sage. Scattered across the ranges are isolated clusters of dwellings, including the ubiquitous *hogans* (the earth-covered, wooden structures typical of the Navajo). Pick-up trucks or horse-drawn "buckboards" are used for transport, although there is the occasional lone rider on horseback. Small settlements are built around the essential and often crowded trading post. In the larger towns, identical prefabricated homes focus around the water tower, school, hospital or administrative building. There are alarming levels of unemployment and signs of rural poverty abound. Arizona alone has 23 nominally sovereign reservations, which incorporate 80,000 square kilometres (30,000 square miles) of land, and house about 80 per cent of the state's 200,000 Native Americans. However, reservation areas are largely economically marginal lands.

Left *The setting Sun casts long shadows on low hills at Anza Borrego Desert State Park, California, USA. The beauty of the desert is held in high regard by many Californians. However, many others seem blind to the beauty, and regard the desert as a wasteland, suitable only for cross-country motor sport, which damages the desert's ecosystem.*

Above *Elderly golfers play on artificial grass at Sun City West, a desert resort/retirement centre. Many such communities were developed during the 1970s and 1980s. In part, this represents an overall population migration to the so-called Sun Belt (Florida was also affected); however, it also reflects to some degree the increased life-span of affluent Americans.*

COASTAL CHILE AND PERU

The Pacific coast owes its aridity to the Humboldt Current, which brings cold water from the Antarctic, cooling the surface of the ocean and producing fog and stratus clouds, but almost no rain. The Atacama Desert of Chile and the Sechura Desert of Peru are both hyper-arid. In the Peruvian desert, direct contact with the sea produces a cloudy climate, with nocturnal fogs. The Atacama lies on mountains and intervening basins parallel to the coast, with marked differences in the degree of oceanic influence.

Basins and pans

The Atacama, between Arica and Vallenar, occupies two longitudinal belts: the Intermontane Longitudinal Depression and the Coastal Cordillera (or Tange); but from Chañaral to the southern limit, it is developed in transverse basins. The Intermontane Longitudinal Depression is divided into three parts: between Arica and Quebrada de Tana it is a plain, or piedmont, of sediments accumulated at the foot of the Andes. Near the Coastal Range, these sediments form gently sloping *playas* (saline deposits), which are deeply incised by valleys known as *quebradas*, whose rivers drain into the sea but are very intermittent. From Quebrada de Tana to the Loa River, the depression is closed and contains the Pampa del Tamarugal. Finally, between the Loa River and the basin of Chañaral, there are well-defined basins without runoff of water, but open to the sea. In these are found closed saline depressions known as *salars*. A typical feature of the contact between the inner foot of the Coastal Range and the Longitudinal Depression is the veneer of sodium nitrate, deposited by the evaporation of ancient lakes.

Desert climate

There are two types of climate in the Atacama. The narrow coastal belt has nearly 110 cloudy days in the year, mainly in winter. Stratus clouds and fogs known as *camanchacas* are typical. Nevertheless, the lack of rainfall is almost absolute, and comes only from sporadic, severe rainstorms, often many years apart. The fogs support sparse grass cover in the north and *cactaceae* (cactus) steppe in the south. In this landscape, named *lomas*, mammals such as the two species of fox, a vampire bat and two species of *Pinnipedia* are found. The droppings of marine birds, known as *guano*, are still gathered from offshore reefs and islands for use as fertilizer.

It is in the intermontane depression that the desert is at its most severe, with great atmospheric clarity, low humidity, large daily variations in temperature and an almost complete absence of rainfall. The Atacama is a temperate desert, because mean temperatures are less than 18°C (64°F). Vegetation is absent in these plains, apart from the *quebradas* and the Pampa del Tamarugal.

The Pampa del Tamarugal is a closed depression, in whose lower parts are *salars* produced by occasional floods derived from groundwater. In the eastern piedmont, groundwater allows patches of *tamarugos* (*Prosopis tamarugo*) to flourish. The tamarugo was heavily cut during the heyday of nitrate mining, but is now being replanted in the Salar de Pintados in order to develop the area for the breeding of sheep.

Desert
Semi-arid
Main roads
Railways
Rivers
Seasonal rivers
Salt flats
◆ Mining/mineral exploitation
● Large cities
○ Important desert towns/settlements
International boundaries

0km 500
0miles 250
250

N

Talara
Piura
SECHURA DESERT
Chiclayo
Trujillo
Chimbote

A N D E S

PERU

Callao LIMA
Pisco

PACIFIC

OCEAN

C
O
R
D
I
L
L
E
R
A

Arequipa LA PAZ

BOLIVIA

Arica

A T A C A M A   D E S E R T

Salar de Uyuni

Iquique

Loa

Tocopilla Chuquicamata
Calama

Tropic of Capricorn

Antofagasta Salar de Atacama

CHILE

ARGENTINA

Chañaral
Copiapó

Vallenar

Coquimbo

Valparaíso
SANTIAGO

Right *The west-coast desert of South America lies between latitudes 5° and 30° S , occupying a narrow strip along the coastline of Peru and extending into northern Chile. The product of cold offshore waters and constant high pressure, the desert area is sandwiched between the Pacific Ocean to the west and the Andes to the east.*

Left *A satellite image showing part of the Salar de Atacama saltpan to the east of the Domeyko Mountains in northern Chile. Much of the Atacama Desert occupies a series of high basins, or troughs, located between barren peaks. In these inhospitable inland regions, the average elevation of the desert landscape is in excess of 2,000 metres (6,500 feet).*

DEVELOPMENT OF THE ATACAMA

Human settlement and agriculture in the Atacama have been possible only around four groups of oases: Arica near the Peruvian border, the eastern edge of the Pampa del Tamarugal, and the basins of the Loa and Copiapó rivers. When the Spanish arrived, they found that the valley bottoms, or *quebradas*, in the longitudinal depression between Arica and Quebrada de Tana were extraordinarily fertile. The chronicler Cristóbal de Molina described the area's "beautiful farmlands" and irrigation systems. According to the natives, it never rained but the area never lacked water. In each valley "thermal rivers" or spring waters allowed them to irrigate their lands and vegetable gardens. Today, irrigation in this region is possible only by means of wells, whose water is generally slightly saline. In modern times, the principal crops are corn and alfalfa.

For the villages along the eastern edge of the Pampa del Tamarugal, irrigation is made possible by traditional subterranean conduits similar to the *foggara* of the Sahara. In the central Atacama, the Loa River – whose source lies high in the Andes – has been used to irrigate alfalfa. The Loa waters Calama, the ancient centre for the production of nitrates as well as the modern centre of Chile's copper industry. Alfalfa is also grown in the valley of the Copiapó River and in the Vallenar Basin in the Norte Chico semi-desert. In the Copiapó valley in particular, agricultural modernization has produced major changes since the 1980s. These include new patterns of land use, the expansion of land under cultivation and the increasing production of fruit.

Mining the Atacama

Following the discovery of silver by Spanish pioneer Juan Godoy near Copiapó in 1832, mines were opened up in the Atacama. However, until the construction of a railway between Copiapó and Caldera Bay – the first railway in the Southern Hemisphere – exports were limited by the capacity of mule trains. Later, silver was discovered near Chañarcillo, and the population of the Copiapó valley grew spectacularly.

Despite the silver rush of the 1800s, mineral exploitation of the Atacama dates back to pre-Columbian times, when farmers along the eastern border of the Pampa del Tamarugal extracted *caliche*, a salt deposit containing sodium nitrate (saltpetre), used in the manufacture of fertilizer. After the arrival of the Spanish, the crushing and leaching of *caliche* flourished as a home industry.

During the 1800s, Chilean nitrates were heavily exploited, following the exhaustion of the guano deposits along the coast and on many offshore islands. The nitrate industry boomed toward the end of the century. Rail construction made possible the development of several centres of production. At the industry's height, 150 factories were in operation. But after the German invention of synthetic nitrates during World War I, the Chilean nitrate industry suffered severely, and the landscape of the nitrate *pampas* began to change. Only a few factories are active today, and decayed remnants of towns, villages and railways still dot the desert. Nevertheless, present production still provides nitrate, iodine, borax and sodium sulphate, which are used for various purposes.

146

Left *Pouring copper ingots at Chuquicamata, Chile, the site of the world's largest copper mine. Although most desert environments are still relatively free of industrial pollution, the presence of heavy industry has had a damaging effect on parts of the Atacama. In addition to fumes given off during the refining process, huge amounts of dust are created by the open-cast mining techniques used for extracting the copper ore.*

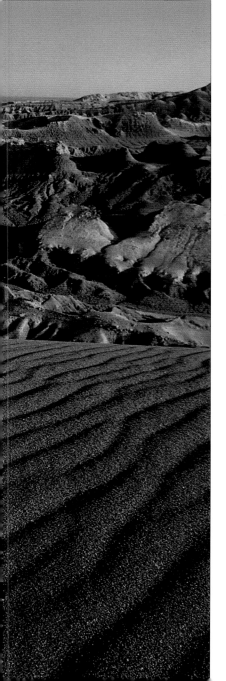

Above *Pisagua, a nitrate port at the edge of the Atacama Desert on Chile's northern coast. Although settled since colonial times, Pisagua earned prosperity from the nitrate railway which runs inland to deposits in the desert. During the heyday of the 1930s, the town was rich; today it has an atmosphere of decay and neglect.*

Left *The Atacama Desert to the southeast of Chuquicamata, Chile. Although parts of the Atacama receive sufficient rainfall for there to be obvious signs of erosion by surface water; the desiccating effects of the Sun and strong winds render it one of the most inhospitable places on Earth. Vegetation is entirely absent, in part due to the high concentrations of minerals in the underlying soil.*

The mining of copper

Copper mining in the Atacama dates from the 1700s. With the decline of nitrate mining, copper increased in importance. On the western ridge of the Andean Cordillera, ancient volcanic activity and seismic movements produced mineral-bearing strata, including the world's largest concentration of copper ore. The largest mines were established after 1915, mainly near the Loa River and Potrerillos, in the Chañaral basin. At Chuquicamata, terraces have been blasted ever downward, so that the mine is now some 580 metres (1,900 feet) below the surface of the desert.

From 1938 to 1975, over 220 million tonnes (tons) of tailings from the Potrerillos and El Salvador copper mines were dumped on the coast, in Chañaral Bay, causing dramatic changes in the beach, which advanced at a rate of some 25 metres (80 feet) each year. The contaminated area extends for 16 kilometres (10 miles) along the shore, and sandy beaches now have a high content of tailings. A dramatic drop in biological diversity in the area seems to be associated with copper in the sediments and pollutants in the water.

THE ANDES AND PATAGONIA

In southern Peru, the Andes Mountains divide into two major prongs: the Cordillera Occidental and the Cordillera Central. These high ranges enclose the desert of the Altiplano (high plains), which lies mainly within the borders of Bolivia. The Altiplano is a closed depression, 3,500–4,000 metres (11,500–13,000 feet) above sea level. It is filled by eroded sediments and volcanic material. The region is characterized by, among other things, vast flat *salars* (salt basins), which are the remnants of ancient lakes.

In Chile, the Codillera de Domeyko – an offshoot of the Cordillera Occidental – encloses the Salar de Atacama in an intermontane depression. The basin is filled with sediments, and a broad, elongated *salar* at 2,300–2,400 metres (7,550–7,850 feet) above sea level. In the oases along the eastern edge of this basin, farmers raise crops and livestock.

South of the Bolivian Altiplano, northern Argentina contains a number of inward-draining depressions. The principal of these depressions contain *salars*. The southern end of the Andean desert is where the Cordillera Central and Cordillera Occidental converge.

Climate, flora and fauna

Climatic conditions on the Altiplano are marked by strong and persistent winds and great daily temperature variations. In the Bolivian section of the plain, annual rainfall averages some 200 millimetres (8 inches), with no marked seasonal variations. However, rainfall drops significantly from west to east. Visitors to these high regions almost invariably suffer from altitude sickness.

At an altitude of around 3,500–4,000 metres (11,500–13,000 feet), the vegetation consists mainly of small shrubs. The dominant species is utola (*Baccharis tola*). On the western slopes of the Cordillera Occidental, the principal plants are various species of Cactaceae (cacti and their relatives). On the slopes of the salars in the Puna, the gramineous (grassy) species known as *pajonal* dominates. In the basin of the Salar de Atacama, slightly higher rainfall allows the development of some denser plant communities. The chief species is the cachiyuyo (*Acantholippia atacamensis*), which is consumed by cattle in the oases. The lower slopes of volcanoes bordering the basins of the Altiplano are home to llareta plants (*Azorella compacta*), which are used for fuel.

The semi-desert of Patagonia

In the southern Argentinian provinces of Chubut and Santa Cruz, arid tablelands dominate the landscape from sea level to about 1,000 metres (3,300 feet). Because the prevailing westerly winds lose their moisture while crossing the southern stretches of the Andes, Patagonia is a cold semi-desert in which precipitation falls principally in winter. Temperatures average only 7°C (45°F) throughout the year, and the many days of frost and snow, and persistent westerly winds restrict the vegetation cover considerably. One of the most serious problems in Patagonia is erosion of the soil; wind-borne dust damages the sparse vegetation. In closed depressions known locally as *bajos*, the accumulation of dust and salts produce *salitrales* (salt pans).

Below *Satellite image showing Lake Poopó in the Bolivian Antiplano. No more than 3 m (10 ft) deep, the lake covers about 3,000 sq km (1,150 sq miles) at its normal extent. In exceptional circumstances, the waters reach to the town of Oruro, nearly 50 km (30 miles) distant from the usual shoreline. Lake Poopó lies at the lower end of a surface drainage system which starts in the north with Lake Titicaca. Lake Poopó drains into the Salar de Coipasa.*

Right *The high-altitude desert of South America stretches almost the full length of the continent, from equatorial Colombia in the north to Tierra del Fuego in the south. The desert reaches its greatest width in the Antiplano of Bolivia. The windswept plains of Patagonia occupy most of the lowland to the south of the Rio Negro.*

Cali

BOGOTA

COLOMBIA

Equator

0km   1000

0miles   500

ECUADOR

QUITO
Ambato
6,267m

Guayaquil

Cuenca

Fortaleza

N

Chiclayo

Trujillo

Chimbote

Huánuco

Pucallpa

PERU

Natal

SERTÃO

Recife

Huancayo

LIMA

Cusco

BRAZIL

Salvador

Arequipa

Lake
Titicaca

LA PAZ

Cochabamba

6,520m

Oruro

Santa Cruz

Arica

Lake
Poopó

BOLIVIA

SUCRE

Salar
de
Uyuni

Potosí

Tropic of Capricorn

PARAGUAY

Antofagasta

Salar de
Atacama

Salta

CHILE

San Miguel
De Tucumán

Santiago del Estero

Salado

La Rioja

Salinas
Grandes

6,323m

Córdoba

San Juan

6,959m

Valparaíso

Mendoza

Rosario

SANTIAGO

Rio Cuarto

URUGUAY

San Rafael

BUENOS AIRES

Concepción

ARGENTINA

Santa Rosa

SOUTH

ATLANTIC

OCEAN

Colorado

Salado

Valdivia

Neuquén

Negro

Bahía Blanca

Puerto Montt

-40m

Valdés
Peninsula

Chubat

Rawson

Chico

High mountain desert

Semi-arid

Main roads

Railways

Rivers

Salt flats

Comodoro Rivadavia

Mining/mineral
exploitation

Deseado

Puerto Deseado

Oil and gas fields

Oil pipelines

Large cities

Río Gallegos

Important desert
towns/settlements

Punta Arenas

International boundaries

149

PLATEAUS AND PLAINS

Before the Spanish Conquest, Lake Titicaca in the Altiplano was holy to the Incas. The Spanish began to occupy the Altiplano in the late 1530s, and the Conquest was more or less complete by 1600. The Spanish presence in the Altiplano region has never been large, however, when compared to other areas of South America. As a result, a large part of the population is still purely or largely of aboriginal descent.

The northern Altiplano receives enough rain to grow grains such as barley, wheat and oats, which are often produced as cash crops. Small farmers, however, who cultivate the majority of the land, generally grow subsistence crops, such as potatoes, quinoa (a grain used in many ways) and canagua, which is related to quinoa.

The living conditions of the Altiplano peasants are very basic. Small adobe huts are typical, and the majority of farms consist of only a few hectares of poor land. Most farming continues without the use of machinery or even animals. The few domestic animals that are kept include llamas, alpacas and chickens. Sheep are the more important domestic animal in the southern parts of the region.

There are few resources to exploit. Exceptions are the very fine wool taken from vicuña and guanaco skins. The animals are hunted in the high grassland areas. Vicuña in particular have come under severe threat from hunting. In the late 1960s, the population was estimated to be below 10,000. Hunting was banned for 10 years in 1969, and it is hoped that the creation of national parks will save the animals.

Mining the Altiplano
The Spanish began to exploit the mineral wealth of the Altiplano region in the 1500s. The city of Potosí, for example, started to grow rapidly when silver was discovered in 1545. Tin mining started to become economically important towards the end of the 1800s.

The tin is often found in the same regions of the Altiplano as silver, a belt that runs from north to south through the Cordillera Real. Both tin and silver are still mined in large quantities. There are also mines producing smaller amounts of copper, lead, antimony, tungsten and zinc.

The environmental damage caused by mining, particularly large open-cast pits, is predictable, if usually fairly localized. However, the damage to a particular region can be devastating. Economic problems arising from falling world commodity prices may not directly increase such problems, but serve to prevent many clean-up and damage-limitation operations that might have been undertaken.

Farming and mining in Patagonia
Until the late 1800s, Patagonia was inhabited by only a few, widely dispersed Indians, most of whom lived by hunting. In the 1880s, Argentinian troops undertook a campaign that gained control of the region. There are now few descendants of the original inhabitants living in Patagonia, and there has been no very large immigration of Europeans. After gaining control of the region, the Argentine government decided to give away large blocks of land, and most of these are now run on the estancia ranching system.

Most agriculture consists of livestock rearing in the vast dry areas, although some irrigated parts, such as the upper Negro Valley, grow a wide variety of fruit and some vegetable produce. Because of the harsh environment, the shelter provided by valleys is as important as a supply of water.

Minerals extracted include coal, gas and oil, the latter two being produced particularly in the Comodoro Rivadavia region. There are significant deposits of iron ore in Sierra Grande, but only small amounts of other minerals, including uranium, lead and zinc.

Large hydroelectric schemes have been built on the upper branches of the Rio Negro. These schemes seem to have caused no severe ecological damage other than the flooding of quite large areas of the desert. More schemes are planned, however, and it is hard to tell what the effects might be.

The main ecological problem arising from human activity is the destruction of vegetation by the more or less uncontrolled grazing of animals. The increased soil erosion that this causes has led to considerable expansion of desert areas.

Above A high, cold and very dry desert in the Puna de Atacama in northern Argentina. This area suffers all the vicissitudes of deserts and of cold climates, such as wind erosion, dust storms, and frost-shattering of rock. The boulders were probably dumped by glaciers in the not too distant past.

Left Market square in Tarabuco, Bolivia, situated above the Antiplano, on the slopes of the Eastern Cordillera. Mountain rainfall, supplemented by spring water, enables the inhabitants of the district to grow a wide range of fruit and vegetables. The blurring of the seasons in this tropical region means that fresh food is available to the people all year round.

DESERTS OF AUSTRALIA

Australia has few totally barren deserts, and yet, except for Antarctica, it is the most desertified continent of all. This apparent paradox results from the fact that a few thousand years ago Australia was more arid than it is today and has inherited many desert-like landforms. In addition, rainfall is so unreliable that areas in the semi-arid zone may appear very arid at times. However, overall there is sufficient rain for the survival of hardy vegetation.

In Australia, the arid zone lies between the 250-millimetre (10-inch) isohyet (a line drawn through areas that share the same amount of rainfall) in the south, and the 500-millimetre (20-inch) isohyet in the north. The Australian climate is notable for great variability from year to year.

Desert types

Australia's arid zone can be divided into three main types; clay plain deserts, sandy deserts and stony deserts. Clay plain deserts are fundamentally alluvial plains formed when rivers deposit clay and silt. Sandy deserts are characterized by remarkably parallel linear dunes. The Simpson Desert, an example of this type, has exceptionally long and straight dunes, between which lie clay swales (moist depressions). Sandy deserts with lakes may also have *lunettes*, crescent-shaped dunes on the downwind side of lakes.

Stony deserts can be subdivided into four smaller categories – shield deserts, range and valley deserts, limestone deserts and gibber plains. Shield deserts, sitting on the Precambrian Shield, an area of ancient granite and metamorphic rocks, dominate Western Australia. Much of this shield is deeply weathered and has a long landscaping history.

Range and valley deserts are deserts that simply lie on folded strata, while limestone deserts, such as Nullabor Plain, lie on horizontal limestone. Gibber plains are stony deserts paved with many stones (gibbers or desert pavement) that overlie fairly stone-free soil.

Much of the drainage of inland Australia is internal, and surface runoff is unreliable. Groundwater is often saline, but in some situations a thin layer of fresh water lies on top. This layer is important for some species of plants and animals.

Australia's desert lakes range from very small to vast. The largest is Lake Eyre, which lies about 12 metres (40 feet) below sea level and has an area of 9,300 square kilometres (3,600 square miles). It filled for the first recorded time in 1950 and took two years to dry out.

Geological history

When the ancient supercontinent of Gondwanaland broke up, the final split between Australia and Antarctica occurred about 55 million years ago. Antarctica has hardly moved since then, whereas Australia has drifted north. The continent did not drift from a cold climate to a warm one, but was warm and wet until it reached its present position. Drying started a few million years ago, but the exceptionally arid conditions of modern-day Australia started less than one million years ago. Although one million years is a short period in geological timescale, the dry conditions have existed long enough for biological adaptations to emerge.

Below *A satellite photograph showing the vast stretches of sand dunes around Lake Torrens in South Australia. Earth movements have created a series of depressions in the desert landscape, which occasionally fill with water after heavy rain, creating temporary lakes. Even when full (as shown) the surface of Lake Torrens lies some 25 m (80 ft) below sea level.*

Right *Australia is the most arid of the continents, and only 10 per cent of the land (confined to the northern and eastern coasts) receives more than 100 cm (4 in) of rain per year. The most arid region is in the centre, Australia's "Red Heart", where a scant 10 cm (0.4 in) of annual rainfall barely supports clumps of porcupine grass among the sand dunes. Around the dunelands are areas of dry savanna where dwarf eucalyptus, acacia and saltbush (Atriplex spp.) grow.*

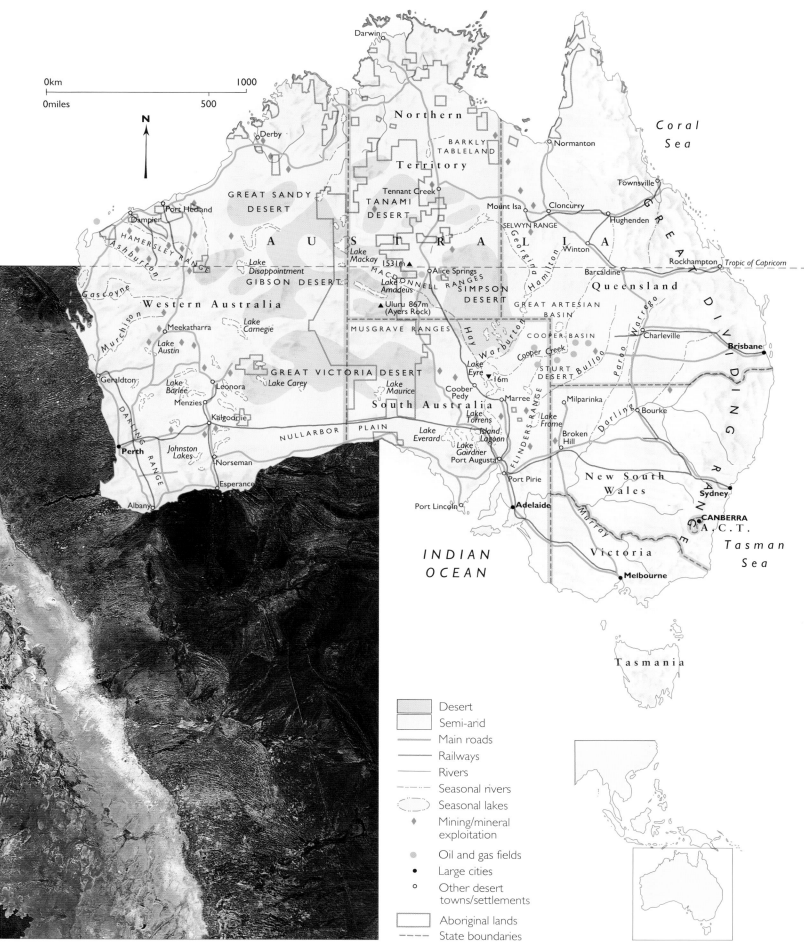

Desert

Semi-arid

Main roads

Railways

Rivers

Seasonal rivers

Seasonal lakes

Mining/mineral
exploitation

Oil and gas fields

Large cities

Other desert
towns/settlements

Aboriginal lands

State boundaries

153

PEOPLE IN ARID AUSTRALIA

People have survived in the Australian arid zone for many thousands of years. The earliest human remains are about 40,000 years old and belong to ancestors of the Australian Aborigines. Occupation sites, however, have been discovered on the continent which date back about 60,000 years, and biological evidence suggests that people may well have arrived even earlier – about 140,000 years ago.

When European settlement began in 1788 the numbers of Aborigines declined very quickly, especially in the fertile margins of the country. The Europeans grouped them together and placed them in missions and reserves. Although they now have their own tribal lands, many have moved into the towns. In recent years, however, there has been an "outstation" movement, with small groups of Aboriginals moving to remote settlements, where they can live in a more traditional manner in the desert.

Aboriginal hunter-gatherers are generally perceived as living in harmony with their environment. Recently, some scientists have suggested that their hunting affected the local animal populations, and that larger animals may have been killed off. However, there is little evidence to support this. It has also been suggested that Aboriginal gathering of plants for food affected the Australian ecology.

The European invasion

Whether Aboriginal hunting-gathering techniques damaged the environment or not, there is little doubt about the effect Europeans had on the arid and semi-arid lands. The consequences of clearing, overgrazing and over cultivating were devastating. Massive clearing projects were undertaken to make way for the pastoralists and cultivators. They would move into the arid lands during a "good" time, when the rains were heavy, but then withdraw when the land once again became too arid, leaving a barren wasteland behind them.

Wheat was, and still is, the dominant crop, and the margins of the wheat belt have moved by hundreds of kilometres (miles) over a few years. The arid zone in South Australia and New South Wales, for example, moved up to 200 kilometres (125 miles) towards wetter areas from 1881 to 1910 and 1911 to 1940. Following the drought of the 1960s, a 300-millimetre (12-inch) annual rainfall line was defined as the minimal safe margin for growing wheat. Drier areas, however, were developed for wheat growing during a spell of wetter years with the result that following the drought of the 1980s, whole settlements turned into ghost towns.

Clearing forest and bush was done for several reasons other than making the ground available for cultivation. In the Kalgoorlie region, for example, tree-stumps show that this "arid" area was forested at the time of the first gold rush in the late 1800s. The trees were destroyed in vast numbers for building and for smelting gold.

On marginal lands, the extreme technique of "recreational ploughing" is practised. If good rains follow, a crop is achieved; if the rain, however, is insufficient, the farmers still collect a subsidy. Such practices contribute greatly to erosion. As a result, dust storms have been known to carry "red rain" from the topsoil of the arid zone as far as Melbourne.

Livestock management

To compound the problem of erosion, many cattle and sheep have been reared in the arid lands. Their numbers often exceed the carrying capacity of the land, leading to severe erosion. In western New South Wales, 15 million sheep grazed on the arid lands before the drought of the 1890s, when the number was reduced to 3 million, returning later to 7 million. Many people believe that trampling by hoofed animals was particularly destructive in Australia, where the native animals have soft feet.

Introduced animals also compete with the native fauna for food and water. This applies not only to domestic animals, but also to their feral equivalents (wild horses, camels, buffalo), and to the rabbit and its introduced predator, the fox. Introduced plants also change the ecology. Buffel grass, for example, was introduced to conserve soil and halt erosion. However, it is inedible to most indigenous animals.

A modern form of land use in the desert is tourism. A few places have become major attractions, and Uluru (formerly Ayers Rock) now has over 250,000 visitors per year. Such numbers require considerable development of roads, water supplies, waste disposal and accommodation, all of which have a marked effect on the environment.

Left Cattle being mustered up at a station (ranch) in Western Australia. After rainfall, the land becomes green with grass, but this is soon burnt brown by the Sun. In such a climate, cattle must be ranched extensively, with less than one animal per square kilometre (three per 2 square miles). Mustering the herd over great distances brings its own problems.

Below Inspecting a dingo fence after a rainstorm. Livestock introduced from Europe is vulnerable to the dingo, the Australian wild dog. In order to protect the main sheep-farming area, the Australian government built a 5,000-km (3,000-mile) fence. The fence is 1.8 m (6 ft) high, and the mesh extends some 30 cm (12 in) below the ground.

Left Aborigines near Alice Springs in the Northern Territory, Australia, burning traditional designs into a boomerang for sale to tourists. The widespread availability of rifles has rendered the boomerang largely a conversation-piece or a toy, even among most Aborigines. Only a few of the older hunters still practise some of the traditional skills of the pre-colonial era.

155

WATER AND MINING

Because surface water is scarce in Australia, groundwater or imported water is used. Much of eastern Australia lies on the Great Artesian Basin, a vast, basin-shaped aquifer that has an area of some 17 million square kilometres (6.5 million square miles), about 20 per cent of Australia. The Great Artesian Basin leaks naturally in a few areas, but there are about 18,000 boreholes tapping the resource. Despite exploitation the basin is still virtually full, and being recharged. Elsewhere, however, where groundwater deposits are being mined, recharge will not replenish them.

Watering schemes

Some towns are maintained by piped water. Kalgoorlie in Western Australia, for example, was supplied by the Goldfields Water Scheme, which was completed in 1903. The scheme consisted of a 520-kilometre (320-mile) pipeline conveying over 20 million litres (4.4 million gallons) per day. This has since been extended to the Goldfields and Agricultural Water Scheme that now serves 30,000 square kilometres (12,000 square miles) of farmland.

Australia has both active irrigation schemes and the rather more fanciful "engineer's dreams". The economics of irrigation have been severely attacked, but the practice continues. Worse is salination. The Murray Valley irrigation area, for example, is losing land at an alarming rate due to rising water tables, caused in turn by irrigation.

The best known of the so-called "engineer's dreams" was the Bradfield scheme of the 1930s, whereby the Tully, Herbert and Burdekin rivers were to be diverted into an inland irrigation area. Another scheme was to flood Lake Torrens in South Australia with seawater and so affect the climate of the surrounding area. These schemes now seem impractical and uneconomic, even without consideration of the resulting salinization.

Above *A pipe from an artesian well supplies a cattle trough in the Northern Territory, Australia. While the use of artesian water is essential to cattle ranching in many parts of Australia, its misuse is becoming increasingly hard to justify and risks the sustainability of the resource. The trough here is brimming, the ground is saturated with the overflow, and yet there is not an animal within sight.*

Right *An opal miner picks his solitary way to riches beneath Coober Pedy, South Australia. Opals are an apt metaphor for the riches of the Australian desert. Each stone owes its delicate coloration to the presence of minute quantities of water. If an opal is allowed to dry out completely, it loses its colour, and its value (in human terms) declines enormously.*

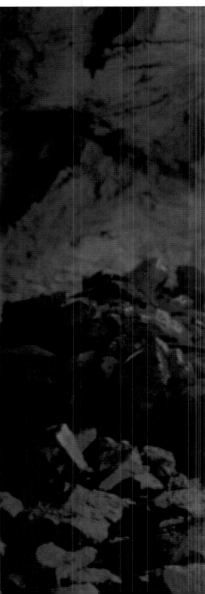

Mining the desert

In Australia, the only economic mineral that is directly associated with arid conditions is opal. Opal is formed by the precipitation of silica-bearing solutions near the Earth's surface, which combine with salt from salt lakes. Coober Pedy in South Australia is a famous opal town, which has changed from a tiny settlement where the miners actually lived underground (to escape the heat) to a developed town with a water supply and many surface buildings.

Other mineral deposits have been found in the desert. These include gold, with major deposits in the Kalgoorlie area; lead and zinc, with major deposits at Broken Hill; copper and uranium at Olympic Dam; and iron ore at Mount Tom Price. These sites typify the large-scale mineral finds, and demonstrate all the problems encountered by remote towns in deserts, such as water supply, communication, entertainment, education and social relations.

Some mines are underground, but many are large open-pit excavations. To the effects of mining should be added the disturbance caused by exploration (often by networks of vehicle tracks), and waste disposal. The mines have resulted in the establishment of an infrastructure of railways, roads and pipelines. Mining covers about half of one per cent of Australia's land area, so its environmental impact is much less than agricultural or pastoral activities, or even urbanization. However, the massive clearing programmes that accompanied the gold rushes of the late 1800s have left behind them vast barren tracts of land.

The potential environmental damage caused by various mining activities has forced all major mining companies to employ their own environmental officers. They are concerned with all aspects of mining, pollution, waste disposal and reclamation. Mining companies face extremely close scrutiny and are expected to maintain high standards of environmental management. Both the federal and state governments have enacted legislation requiring the assessment of the environmental impact of new mining projects.

Another issue raised by mining operations is that of Aboriginal landrights. Only recently have Australian Aborigines been granted rights to land that is important to them for cultural and historical reasons. In the past, the discovery of potentially valuable mineral deposits in Aboriginal lands has resulted in massive relocation programmes, often from culturally significant sites.

THE CHALLENGE OF CONSERVATION

Desert environments are highly sensitive and extremely susceptible to the effects of new usage. Local threats from mining, industry and overgrazing are problem enough, but major changes in the use of water and possible climatic change can cause even greater damage. What little life there is in deserts can be all too easily destroyed, and desert margins may become increasingly desertified as human activity ruins productive land. While some of these problems are global and must be addressed as such, others relate specifically to the use of particular desert areas. It is essential to formulate strategies for restricting damage and for ensuring the sustainable use of these harsh yet delicate environments.

Above *Furrows in Australia give vegetation a foothold.*
Right *A fence in Australia marks the boundary of overgrazed land.*

DESERTS ON THE BRINK

The threat of desertification is real. As has been described earlier, it occurs when ecological processes and conditions characteristic of deserts extend into formerly semi-arid areas. It is possible for this to happen in two major ways.

First, desertification can occur when there are major changes in climatic conditions. There have been such changes throughout geological history in a quite natural way. The most recent of these has been a steady warming of the Earth since the last ice age, which has in general brought wetter conditions. Today, however, there is the additional factor of the burning of fossil fuels, which is generally considered to be leading to more rapid global warming. This could cause a shift in climatic belts, and possibly the expansion of some deserts, although it also might cause the contraction of others. There is considerable scientific debate over which of these effects might predominate, and what little evidence we have is at best confusing.

Second, it is widely believed that much recent desertification has been the result of human intervention – such as irrigation or overburdening land with livestock – although again there is much debate on the matter. Combined with a series of dry years, some human activities can indeed lead to desert-like conditions. This kind of desertification, however, is not occurring along broad fronts. It is patchy, depending on local ecological conditions and human pressures. As a consequence, initiatives to solve these problems need to be targeted at a local level.

The fragile desert

The ecological systems of the deserts and their margins are delicately adjusted to the local conditions. Besides aridity, these areas experience large variations in temperatures, from the heat of the day to frost at night. They are also subject to extreme events of irregular occurrence, such as floods and fires. Desert ecologies are adapted to long-term variations in rainfall, which typically involve series of years with very low rainfall interspersed with rare years of relatively high rainfall during which the deserts bloom.

Despite this adaptability, and a resilience that allows desert margins to recover from prolonged dry periods, many desert ecosystems are unable to withstand intense human activity. Rapid degradation of soils and natural vegetation may result because marginal systems are constrained by the limited availability of moisture, which restricts plant growth and soil development. Such disruption is difficult to reverse because, similarly, the ecosystems have so few resources for recovery.

Common misconceptions

The conservation needs of desert environments are poorly understood in comparison to issues such as rain-forest depletion. There are two quite common misunderstandings about deserts and their margins. The first is that they are rarely exploited and therefore in a relatively pristine condition. The second is that, because of the low rainfall and apparent barrenness, they are thought to have too few animal and plant species to be of any great importance in the struggle to conserve the world's biodiversity.

The truth is somewhat different. Deserts and their margins are important in the world's ecology for a number of reasons. They are extensive, covering about 30 per cent of the Earth's surface, and diverse, with a variety of biological systems. They are home to a large number of people. Furthermore, they are often exploited.

The pressures that threaten to degrade and destroy desert ecosystems vary from place to place. Some of these pressures are local, such as population growth; others are the result of external pressures as the resources of the regions are exploited. The ability of desert margins to cope with pressures varies. Some human communities have financial resources to invest which can improve the use of natural resources. Other communities, with pressing problems related to poverty, cannot afford the time or money needed to invest in protecting natural resources for long-term use.

It is generally agreed that in recent years desertification has been the result of increased human intervention.

Left In Mauritania, West Africa, a tree maintains a precarious hold on a landscape that is turning to desert around it. The aerial roots are not a desert adaptation, but have occurred as the root system has been progressively exposed by wind erosion of the soil. Lateral root extensions are truncated in favour of plunging tap roots, which bring up moisture from deep below the surface. The open arrangement of the foliage is an adaptation to give a minimum of wind resistance. With so little support remaining from the roots, a structure that tended to catch the wind more might well have been blown over already.

Above Quite the opposite of the tree shown in the picture on the left, these trees in Burkina Faso, West Africa, have been buried by large sand dunes. Even in such a wet-looking landscape, the wind has been able to whip up large amounts of sand and dust from fields, which were exposed during the dry season when crops will not grow. At times when the soil in fields does not have the protection of vegetation – in the form of crops – and becomes dry, it is very prone to erosion. The dunes in the picture bury valuable trees and threaten to bury nearby fields, leading to a complete loss of the productive value of the land.

Left *A farm worker in the drought-stricken Transvaal of South Africa uses a chain-saw to cut leafy branches for starving cattle. Such drastic measures provide only the most temporary, short-term relief. Over any longer period, such actions merely accelerate the degradation of the landscape. Additionally, the investment in time, fuel and expensive machinery involved in this kind of activity is quite disproportionate to the returns.*

Right *Degraded land on the foothills of the Altai mountain range near Samarkand in Uzbekistan. Flocks of sheep and goats, that were banished from the lowlands by new cotton fields, have stripped the grass from these hills, thus leaving them open to attack by running water. It is the pressures caused by development, rather than by traditional land uses, that has brought on most of the degradation. It is often the case that while traditional methods of pastoralism and agriculture complement the environment in which they are carried out, new methods effectively encourage the most potentially harmful aspects of the natural world. Thus, in this case, water has been made an agent of destruction rather than of nourishment.*

For many centuries, communities have survived and prospered in the desert margins, despite periodic droughts and famines. Their success was based on ways of managing resources that were adjusted to the environment or were at very low intensity. They usually survived without destroying or seriously degrading the natural resources on which they relied. In recent decades, the situation has begun to change as a result of population growth, economic pressures, and political and social change.

Population growth

Today, many desert regions face rapidly growing populations. This is partly a result of improved access to modern medical services, which has reduced infant mortality and improved maternal health. It is also a result of the continuation of traditional values that perceive children as a sign of economic strength and source of security in old age. Wealth from oil is another important factor.

In many desert margins, the population is growing because of migration. In some cases, most notably the United States and parts of the Middle East, migrants are moving to towns that have been developed to support industrial and mineral development. Rapid population growth in desert margins, especially in the developing countries, increases the demands upon the regions' natural resources. This frequently leads to degradation, given the limited potential for increased output.

Economic development itself puts ecological pressure on desert regions. Much of this pressure comes from the search for improved incomes by individuals as aspirations rise. This is reinforced by the demands of taxation and declining real prices for agricultural produce, which force farmers to increase their output and use natural resources more heavily in order to maintain their incomes. In addition, many governments frequently encourage the production of crops for export to finance imports or to repay foreign debts.

A number of the ecological problems faced in deserts are caused by the ways in which governments treat indigenous people. Native Americans, the Aborigines of Australia and the nomadic pastoralists of Central Asia and the Sahara have all suffered at the hands of their governments. Their rights have been neglected and land has been taken (usually for development schemes from which they rarely benefit).

Resource conflicts and degradation

Pastoral people rely on their livestock, so as the human population grows, so must the number of animals if standards of living are not to fall. Neighbouring areas in which animals might be grazed are usually also suffering from increased population. The introduction of new crops suited for drier lands leads to agricultural settlement in former grazing areas. Irrigation schemes alienate yet more land for pastoralists, leaving them with less land on which to raise more animals. The lands occupied are often critical to the pastoralists' way of life. They are frequently either the pastures with the highest rainfall or the land flooded during the wet season and it is these areas that provide the only available grazing in the height of the dry season.

The lesson of centuries of sustainable farming has been forgotten in the search for increased incomes.

Social change has led to the collapse of traditional range-management agreements, which in the past prevented overgrazing. Such social changes affect traditional society and reduce the effectiveness of the constraints that this imposed. There is increased competition for resources as populations grow and community structures are disrupted by state intervention. Absentee stock owners are a further problem because they tend to overburden their pastures.

The solutions to these problems vary. One approach is to recognize the rights of pastoral societies, despite their lack of political power, and ensure that their traditional lands are not encroached upon. An integrated approach is required, with attention given to economic diversification, poverty alleviation and increased economic security. In addition, pastoral communities have to re-establish pasture management systems that have in the past and can again ensure sustainable use of the existing resources.

EXPLOITATION AND DEVELOPMENT

Water is the scarcest commodity in the desert. Because demand generally exceeds supply, it is common to tamper with the natural hydrological (water) cycle. Irrigation water, for example, is usually drawn from underground sources that have accumulated over centuries or millennia. Some are recharged by rainfall in neighbouring areas, but others came into existence in wetter times. If these water sources are used more rapidly than they are recharged, the cultivation that they support is unsustainable.

Desert soils contain salts of different kinds. When irrigation is poorly managed, excess water leads to the accumulation of salts in the upper layers of the soil, making the land infertile. Irrigation may also cause waterlogging. The problems of salinization and waterlogging are thought to affect over 30 per cent of the irrigated land in deserts.

Economic change and urbanization
The ecological problems that face agricultural and pastoral communities in desert margins today suggest that there is a need for economic diversification. The changes would reduce the pressures upon the resource base. There are no easy remedies, however, in the alternative economic activities that are being developed. Each has its own ecological problems.

Urbanization is one aspect of economic change and diversification. Urban centres, however, whether in the industrialized countries or in the developing world, place major burdens upon the local environment. In the developing countries, wood is a major demand of urban dwellers who use it to cook their food. So great is the demand for fuelwood that in some countries there is no woody vegetation remaining within a 100-kilometre (60-mile) radius of major urban centres. The need for wood rather than other sources of energy is partly a problem of poverty because, despite the long distances over which the wood must often be carried, it is still the cheapest fuel for the urban poor.

Above In Senegal, West Africa, a vendor rides to market with a load of valuable firewood – "more precious than gold". The metal-framed animal cart, often hailed as an essential tool for developing countries, is here effectively contributing to degradation of the landscape by increasing both the range and carrying capacity of the firewood vendor.

Right A hillside carved away by open-cast copper mining in Arizona, USA. Stripped down to bare rock, the bleak slopes and terraces will take many years to grow even the sparse vegetation seen on the hills in the background. Among the many other forms of environmental damage caused by such mining activity is the creation of large amounts of metal-rich dust that can poison the surrounding landscape.

While fuelwood is not a problem in the desert margins of the more developed countries, water is in very high demand because of the lifestyles of the people in these urban centres. The growth of towns in North America's Sun Belt, in response to the movement of high-technology industries and retired people to the dry and sunny environment, has led to enormous demand for water. In most cases these have been met by using local groundwater supplies and by piping water from outside these regions.

Minerals and cattle

Major mineral deposits are often found in desert regions, for instance, oil in the Middle East, uranium and iron ore in Australia, and diamonds in Botswana. The extraction of these minerals requires water and produces wastes, both of which can damage the fragile desert environment. The search for water to allow mining expansion in Botswana reached the stage where accepting major damage to the world-famous Okavango Delta was considered. The delta provides the critical dry season refuge for the vast wildlife populations that use the Kalahari Desert in the wet season.

> **As much irrigated land has been lost due to poor water management as is presently irrigated in desert margins.**

Economic diversification through modern cattle ranching is affecting the ecology of desert margins, including those of the Kalahari. To encourage commercial ranching, large farms have been established on what used to be communal grazing areas. The resulting pressure upon the remaining communal pasture causes overgrazing. Fences have also been established to keep out game and stop the mixing of cattle and game, which might spread cattle diseases, such as foot-and-mouth disease. In a number of cases these fences have severely disrupted the seasonal movement of game.

The tourist industry

Desert tourism is potentially a more ecologically sensitive form of economic development because it involves little or no alteration of the ecology. In fact, it requires the maintenance of the special ecological characteristics of these areas, especially the wildlife, and so helps maintain biodiversity and ecological systems. Rather than encouraging the expansion of ranching into drier areas, where it is increasingly risky, countries may find that the maintenance of the natural ecology of their deserts offers a better if not more secure return on a smaller investment.

There are nonetheless important questions about how local populations are to derive lasting benefit from the tourism income. It is also essential that the level of tourism that such sensitive areas can both attract and support without damage is carefully conceived and managed.

DESERT CONSERVATION

The future of deserts and their margins gives considerable cause for concern. This is especially true in the developing countries, where population growth and poverty combine to place severe pressures on natural resources. Some people believe that many desert margins are on the verge of ecological collapse. They fear that, even without global warning and other climatic change, land degradation on desert margins will lead to a reduction in the productivity of range and arable lands, the disruption of ecological processes and further extinction of species.

On the other hand, some researchers believe that there is a considerable potential, already seen in some areas, for changes in human behaviour and activities that would prevent catastrophic degradation. The suggestion is that technological innovation and planning interventions can improve the way people use deserts.

The peoples of the deserts are not alone as they face these conservation challenges. At the 1992 United Nations Conference on Environment and Development (UNCED) in Rio de Janeiro there was a call for the development of a "Convention on Desertification". This convention, which would match the one agreed on tropical forests at Rio, would attempt to stress the importance of the environmental problems in the deserts and their margins.

Despite this and other initiatives – such as the United Nations Sudano-Sahelian Office (UNSO), which was set up following the 1968–73 Sahelian drought – progress has been limited. Some projects have been successful, and technical solutions for addressing various problems have been identified and implemented, especially in more prosperous countries. However, sustainable development in most of the desert margins in developing countries has not been achieved, and degradation remains a major problem.

Sustainable development

The World Conservation Union, IUCN, has contributed towards raising awareness of the conservation problems in the desert margins through the "World Conservation Strategy" (1980) and "Caring for the Earth: A Strategy for Sustainable Living" (1991). Addressed to government policy makers, conservation groups, development practitioners, industry and commerce, these documents explain why conservation and development need to be integrated. They show how resource management and the conservation of biodiversity and ecological processes are essential to the future of humankind.

Through these publications and its other activities, IUCN has sought to encourage governments around the world to develop national conservation strategies. These strategies should integrate conservation of natural resources into country-specific development planning to ensure that sustainable development is achieved. The contacts made through these various initiatives have led IUCN increasingly to recognize that ecologically sound management and use of natural resources for the welfare of people has to take precedence over conservation itself. Nonetheless, the maintenance of biodiversity and ecological processes have to be part of the management system.

In order to achieve sustainable development in the desert margins, a number of principles must be followed. These principles, which provide a broad framework for effective action, include the following.

- **Integrated land use management practices**, taking a broad view of resource use, and recognizing how the whole range of land uses interact with each other and need to be managed as a complete system.
- **Institutional development** to ensure local control over natural resources and the build-up of community management capacities.
- **The creation of a favourable policy environment** that would encourage people at all levels to improve the management of the resources that are available.
- **The use of local knowledge,** building on local management techniques and adaptive capacities so that any changes that are made are appropriate to the resources and skills of the communities.
- **A multi-sectoral approach,** recognizing that economic development and diversification are essential to reduce poverty, slow population growth and produce the alternative sources of income that can reduce direct demands upon the desert resource base.

Conservation is about people as well as flora and fauna.

The IUCN believes that communities themselves must be able to take the first steps to address the conservation challenge – to take part in what is known as primary environmental care. Such local action has to build upon improved understanding of both the ecological characteristics of areas and upon the interactions that result when human intervention occurs. It must also be built upon more sensitive actions by governments, both local and external, which must recognize the rights of the communities in these regions. Governments must also act to protect the long-term interests of national and global communities.

Only when the management of desert margins becomes sustainable will it be possible to ensure the survival of national parks and of game and biodiversity reserves within desert areas. These are essential for maintaining the genetic resources of the plants and animals that are so well adapted to desert conditions, and which will be of increasing importance as biotechnology skills improve and we are able to take ever greater advantage of those genetic resources that are available to scientists and researchers.

Conservation is about people as well as natural flora and fauna. It is about achieving sustainable production methods that will ensure the survival of both people and resources. In order that this may come about, a fully integrated view is required of the ecology, economy and society of deserts and their margins. Developing such a perspective is perhaps the greatest challenge of all.

GLOSSARY

Words followed by an asterisk* have entries of their own.

aestivation Seasonal state of torpor during the hot months of the year. Animals undergo aestivation to survive harsh summers for many of the same reasons as some animals undergo hibernation in winter.

alluvial fan Cone-shaped deposit of alluvium* left around the course of a former river.

alluvium Fine sediment deposited by a river. The sediment consists largely of mud, sand and gravel, and is often highly fertile and supports agriculture.

anticyclone Another name for a high pressure zone*.

aquifer Layer of porous rock that contains large quantities of water. The term is usually applied to water deposits that provide water for use by humans.

arid Describing climates or regions with an average annual rainfall of less than 200 millimetres (8 inches).

badlands Desert region that is dissected by dense networks of rills and deep, steep-sided gullies and is devoid of vegetation. Erosion in badlands is rapid.

barchan Another name for a crescent dune*.

biodiversity Number and variety of plant and animal species in a given region.

biome Largest type of ecological community that is generally identifiable as distinct, for example, tundra, savanna, temperate woodland, rain forest or desert.

carrying capacity Maximum number of grazing animals that a given area of pastoral land can sustain without degradation through overgrazing or other problems.

catch crop Product of opportunist crop farming. Such crops are grown only at times when rainfall is sufficient.

continentality Conditions and effects of a continental climate. Such a climate is one that is not affected by maritime influences because it occurs in the central regions of a vast landmass. It lacks moisture and has pronounced differences in temperature between summer and winter.

crescent dune Crescent- or horseshoe-shaped dune. Such dunes are relatively rare but are noted for their speed of movement. (Also known as a barchan.)

deflation Removal of loose surface material by the wind.

desert pavement Angular stones found on the top layer of many desert soils and forming a fairly continuous cover.

desert varnish Dark, surface sheen that is found on the uppermost layers of many desert rock surfaces. Desert varnish is thought to be formed by the action of lichens over long periods of time. (Also known as rock varnish.)

drought Period during which rainfall does not reach expected levels in a given region.

dust devil Small but powerful whirlwind that picks up dust, sand and other loose surface material.

ecosystem Relationships between the members of a community of plants and animals, and the interactions of that community with its physical environment.

environment Physical and biological factors that characterize a certain area with respect to the life in that area, or the set of such factors that affect an organism or species.

enzyme Biological catalyst, that is, a substance in an organism that increases the rate at which chemical reactions take place within the organism. Catalysts, which are proteins, are essential to life because most important reactions would take place far too slowly without them.

erosion Wearing away of rock and the removal of land surface largely by the actions of materials (such as sand) carried by wind or water. Wind and water on their own have relatively little power of erosion, although they carry away already loose material.

evaporite Substance that, having been dissolved in a liquid, is left behind when the liquid evaporates. Salt left on the bottom of a dry lake, for example, is an evaporite.

flash flood Brief torrent of water, usually caused by a heavy rainstorm, that occurs especially in deserts.

floodplain Flat land bordering a river that consists of alluvium* that has been deposited by the river.

fossil water Water that has lain underground for many thousands or millions of years. Such water usually found its way under the ground when the climate of a region was wetter than it is at present.

groundwater Water that lies underground, most of which is found below the watertable.

Hadley cell Circulation of air in which hot, moist air rises at the Equator (where the moisture is released as rain), moves polewards to a latitude of about 30°, and then cools and sinks, flowing back towards the Equator.

high pressure zone Weather system with high atmospheric pressure in which air sinks, diverges outwards and disperses cloud. (Also known as an anticyclone.)

hyper-arid Describing climates or regions with an average annual rainfall of less than 25 millimetres (1 inch).

ice age Any one of a number of periods in the Earth's geological past when glaciers spread to cover much more of the Earth's surface than they do usually. The most recent ice age was the Pleistocene epoch (from about 100,000 to 2 million years ago), sometimes called "the Ice Age."

infrared Part of the electromagnetic spectrum (which includes radio waves, microwaves, visible light and suchlike) of slightly longer wavelength than red light, and invisible to the human eye.

inselberg Isolated, steep-sided hill, usually found protruding from semi-arid plains.

intermontane Area that lies between mountain ranges

isohet Line on a rainfall map that joins together points that have the same rainfall.

linear dune Long, sinuous dune that is fairly straight overall. Such dunes are the most common type and may reach several kilometres (miles) in length. (Also known as a sayf.)

loess Accumulation of dust particles deposited by the wind.

metabolic rate Rate at which the internal chemical processes, and thus many of the functions, of a living organism proceed.

petrochemical industry Any industry involved in the exploitation of crude oil, natural gas and their derivatives. For example, the processing of crude oil into its constituent fractions, or the manufacture of plastics from petroleum.

photosynthesis Process by which plants form carbohydrates from carbon dioxide and water using the energy of sunlight. Plants use the carbohydrates for various purposes, but especially as a source of energy.

potash Common name for a number of potassium-containing compounds. The majority of mined potash is used in the manufacture of fertilizers, although it has many other uses in the chemical industry.

precipitation In meteorology, moisture in the atmosphere condensing and, generally, falling to Earth as rain, dew, hail, sleet or snow.

quanat Artificial underground channel, usually in Iran or Pakistan, for carrying water from a distant aquifer to an area where it is needed for irrigation or other purposes.

rangeland Land that has vegetation suitable for grazing livestock, but that is too arid for crop farming.

renewable resource Any resource that is not diminished when it is used or else replenishes itself very rapidly. For example, the use of sunlight in a photoelectric cell does not diminish the supply of sunlight. On the other hand, resources of oil will, if effect, never replenish themselves.

rock varnish Another name for desert varnish★.

runoff Surface water (usually from rainfall) moving over land, as opposed to being absorbed into the ground.

salinization Increased levels of salt brought about by a variety of processes involving the evaporation of water.

saltation Jumping motion of sand particles that are being moved by the wind.

salt flat/salt pan The residue of salts left behind on the ground when a lake evaporates

sand Fragments of rock with a particle size of between 0.2 and 2 millimetres (0.008 and 0.08 inch). Particles smaller than this are generally referred to as dust, although the distinction is not formal.

sayf Another name for a linear dune★.

semi-arid Describing climates or regions with an average annual rainfall of less than 600 millimetres (24 inches).

stratum Any one of the distinct layers into which sedimentary rocks are divided. Each layer corresponds to a particular period of sedimentation.

succulent Any of a large group of flowering plants with thick, fleshy stems and leaves. The plants are adapted for arid climates and include the cacti.

sustainable Describing a method of agriculture, way of life, power-production process or suchlike that can be maintained indefinitely. Sustainable agriculture, for example, involves such things as farming practices that do not degrade the land, and do not consume non-renewable water supplies.

wadi River or stream that flows only during and following rainfall and is at all other times dry.

water table Level in the ground below which rock strata are saturated with water.

weathering Breakdown of rocks resulting from the actions of such things as wind, rain, temperature change, and plants and other organisms. Weathering does not involve the transport of materials, such as sand carried in the wind (see erosion★).

yardang Elongated hillock of rock that has been formed by the action of the wind.

INDEX

Numbers in *italic* type refer to picture captions; numbers in **bold** type refer to entries in the Glossary.

A

Aborigines, Australian *155*
  clothing 64
  displacement of 59
  earliest remains 154
  land rights 157
  tools and implements 68
  totemic beliefs 70
Abu Dhabi city *76*
Abu Zayd 73
aestivation 52–53, 168
Afar peoples 108
Africa 12, 14
  drought 23
  See also East Africa; North Africa
Agadir 77
agriculture 84–85, *89*
  and settlement 59
  Altiplano and Patagonia 150
  Arabian Peninsula 114
  Central Asia 124, 125
  cults associated with 70
  East Africa 108
  effect on herding activities 82
  Great Plains 134
  Islamic architecture and 62
  oasis 66
  rainfed 85
  Tibet 130
  unsustainable policies 77, 80
Ahaggar Tuareg confederation 106
air conditioning, buildings 60
  Iranian 62
  Phoenix, Arizona 77
Aïr massif 102
  partition of 106, *107*
Alashan Desert 126
Ali-Ilahis 70
alluvial fans 24, 25, 168
  formation of 28
alluvium 168
Altiplano Desert 148, 150
  Lake Poopó *148*
  mineral reserves/extraction 150
Al-Qa'im mineral deposits 95
amphibians 52
Amu Darya floodplain, duststorms 36
Amu Darya river 122, *124*
  effect of damming 124
Anahit, cult of 71
Anasazi people *75*
ancestors, cult of 72
Andes 148
  climate 16
  See also Altiplano
animals, domesticated 83
  Altiplano 150
  materials from 60, 150

animals/wildlife: Aboriginal hunters and 154
  around the Aral Sea 122
  Atacama 144
  effect of oil pollution 117
  Gobi 126
  introduced 154
  Namib 112
  positions in food webs 54
  reduction in populations 112
  water retention 50, 51, 52
anticyclones *14*, 41, 168
Anza Borrego Desert State Park *143*
Apache Indians: clothing 64
  displacement of 59
aqueducts, Colorado River 140, 142
aquifers 21, 168
  depletion of 77
  Great Plains 134
  major Saharan 104
Arab peoples 120
Arabian oryx 50
Arabian peninsula 12, 114, *114–115*
  natural gas extraction 117
  salt marshes 25
Aral Sea 122, *122–123*, *125*
  desiccation of 122, 125
  duststorms 39
  effect of irrigation 125
  water transfer from 86
Archimedes screw 84
Architecture, Islamic 62, *63*, *74*
  mosques 72, *72*, *76*
aridity: defining 16, **168, 169**
Artemisia spp. 136
Aswan 77
Atacama Desert 144, *151*
  cold ocean currents and 14
  development of 146–147
  mineral reserves/extraction 95, 146–147, *147*
  plant types/adaptations 45, *47*
Ataturk Dam 20
Atlas Mountains: groundwater springs 20
Atriplex spp. 136, *152*
Australian Aborigines *See* Aborigines
Australian deserts 12, 152
  animal species 50
  damage to ecosystem 154
  drought 23
  geological history 152
  mineral reserves/extraction 92, 95, 157
  "Red Heart" *152*
  towns/cities 59
  watering schemes 156
Awash River 108
Ayers Rock *See* Uluru
Azawad, partition of 106

B

Bacchus, cult of 71
badlands 132, 168
  erosional 24
  formation of 28
bad-gir 62
bajos 148
barchan 30, *31* **168**
barley 85
Barsuki Desert 122
basins: Atacama 144
  dry lowland 24
  sand seas in 30
bat-eared fox *43*
Bedouin peoples *58, 61, 67*
  and floods *41*
  Wahhabi sect 121
bedrock, erosion of 24
beetles 48
  Namib 112
Benguela Current 110
  effect of 14
Berber peoples 58, 75, 106
  Almoravid movement 121
biodiversity 168
biome 168
birds, adaptations by *53*
blacktail jackrabbit *50*
blue-green bacteria 46
Border Industries Program 135
boreholes 112, 156
Brazil: drought 23
"broken lands" 134
bronchitis 39
Buddhism 130
  in Central Asia 70, 71
building materials 60, 62, 150
  extracting 94
"burning bush" 72
burrows *50*, 51
Bushmen *See* San

C

cacti *44, 45*
  as micro-habitats *55*
  carbon dioxide uptake 44
  protection from the Sun 45
  root systems 45
Calandria discolor 47
Calanscio Serir 102
  oil strike 92
calcretes 34
caliche 146
Caloptropis procera 45
camanchacas 144
camel caravans *7, 128*
camel spiders 48
camels *50, 61, 83, 83*
  as source of materials 69
  Gobi 128
  milk *67*
  temperature regulation 50
canyons 24
"Caring for the Earth: A Strategy for Sustainable Living" 167
carrying capacity 97, 168
Caspian Sea lakes 122

catch-crops 66, 85, **168**
Catholicism 70
cattle 83
cattle ranching *See* livestock rearing
caves 60
centipedes 48
Central Arizona Project 77
Central Asia 12, 122, *122–123*
  customs and religion 70, 71, 72
  main population groups 58, *59*
  mineral reserves/extraction 124
  people and problems 124–125
  temporary shelter in 69
Chalbi Desert 108
chemical alteration: slowness of 34
Chihuahua Desert 132
Chobe River 110
Christianity 70, 71, 120
clay: as building material 60, 62
  pans 132
  plains 12
cliff dewllings, Mesa Verde National Park *75*
climate 16–17, *16–17*
  and desertification 160
  and formation of badlands 28
  artificial 60
  calculating changes in 96
  Altiplano 148
  Arabian peninsula 16
  Atacama 144
  Australia 16, 152
  Central Asia 16
  fluctuations in 16
  Gobi 16
  Himalayas and 130, *131*
  Iranian plateau 16
  Kalahari 16
  Sahara 16, 102
  Sonora 16
Clothing 64, *64*, 106
coastal desert 110
Colorado Basin, water transfer from 86
Colorado Desert 140, *141*
Colorado Plateau *19*, 28, 136
Colorado River system *136*, 140
  Dead Horse Point *29*
  diversion of 77
  irrigation from 140, 142
  water transfer from 140, 142
Colorado sandstone 19
Colorado Springs, flash flood 41
Columbia Plateau 136
communications: duststorms and 39
  Gobi 129
  problems with migrating dunes 38
conjunctivitis 39
conservation 158–167
Continental Intercalaire aquifer 104

MAP SOURCES

UNEP/FAO World and Africa Database
UNEP and Soil Resources, Management and Conservation Service,
Land and Water Development division, FAO, Rome, June 1984

"Remote Sensing of Arid and Semi-Arid Regions"
M. C. Girard, *Nature and Research,* Carnforth 26 (1990) pp.3–9

Natural Vegetation World Map
George Philip, London, 1966

Deserts of the World
M.P. Petrov. pub. Halsted Press (John Wiley and Sons), Chichester, 1976

Les Milieux Naturels Desertiques
Jean Demangeot
Editions C. D. U. et Sedes, Paris 1981

Middle East Economic Digest

"Desertification of Arid Lands"
H. E. Dregne
Advances in Desert and Arid Land Technology and Development, Volume 3
Harwood Academic Publishers, London 1983

Atlas Mira
Moscow 1974

Oil Industry
International Petroleum Encyclopaedia
Pennwell Publishing, Tulsa 1989 ed.

Counter Risks Information Services, London

Spelling
The World Book Atlas
World Book, Inc., Chicago, 1986

Times Atlas of the World
Comprehensive 7th ed.
Times Books, London 1987

ARTWORK

14 Richard Lewis; **15** Eugene Fleury; **31**T&C Janos Marfy;
31B Peter Sarsons; **73** Trevor Hill; **85** Mick Saunders; **97** Trevor Hill;
101 Richard Lewis. Thanks also to Vivienne Cherry.

PICTURE CREDITS

(Abbreviations: R = Right, L = Left, T = Top, C = Centre, B = Below)
1 Planet Earth Pictures; **2–3** Oxford Scientific Films/Michael Fogden;
4–5 Oxford Scientific Films/Michael Fogden; **5**TCL Comstock/Richard
Woldendorp/SGC; **5**TR Oxford Scientific Films/M.J. Coe; **5**BCL ZEFA;
5BL Comstock/Richard Woldendorp/SGC; **5**TCR NHPA/Otto Rogge; **5**TL
Oxford Scientific Films/Michael Fogden; **5**BL Aspect Picture Library/
Antoinette Jaunet; **6–7** Magnum/© Steve McCurry; **8–9**T Oxford Scientific
Films/Richard Packwood; **10–11** Geoscience Features; **11**TR Comstock/
Richard Woldendorp/SGC; **12–3**BR Comstock/Anthony Howarth/SGC;
13CR ZEFA; **13**T Aspect Picture Library/Phil Conrad; **15** Shell Photo
Service; **16–7**T Aspect Picture Library/Peter Carmichael; **17**TC inset NHPA/
Anthony Bannister; **17**BL Comstock/Nick Holland/SGC; **18–9**T Bruce
Coleman Ltd/Jeff Foott Productions; **19**C inset Oxford Scientific Films/
J.A.L. Cooke; **19**BR Planet Earth Pictures/John Lythgoe; **20–1** Science Photo
Library/David Parker; **21**CR Hutchison Library/Crispin Hughes; **22–3**
Magnum/© Sebastiao Salgado; **22–3** Planet Earth Pictures/Hans Christian
Heap; **24**TL NHPA/Otto Rogge; **24–5** Bruce Coleman Ltd/Steven Kaufman;
26LC Planet Earth Pictures/David Jesse McChesney; **26–7** Comstock/Alastair
Scott/SGC; **27**TR Aspect Picture Library/Peter Carmichael; **28–9** ZEFA;
29BR NHPA/Grant Dixon; **29**BL Robert Harding Picture Library; **30**TL
inset ZEFA; **30–1**T ZEFA; **31**TL inset Oxford Scientific Films/Richard
Packwood; **32–3**T Hutchison Library/Tuck Goh; **33**BL Andrew Warren;
34–5 NHPA/Anthony Bannister; **35**CL inset Andrew Warren; **35**CR Science
Photo Library/Martin Land; **35**C Andrew Warren; **36–7**B Oxford Scientific
Films/Andy Park; **37**TL Bruce Coleman Ltd; **38**BL Magnum/© Steve
McCurry; **38–9** ZEFA; **39**TR inset Hutchison Library; **40**BR Aspect Picture
Library/Peter Carmichael; **40–1** Frank Spooner Pictures; **42–3** ZEFA; **43**TR
Oxford Scientific Films/M.J. Coe; **44–5** Planet Earth Pictures/William M.
Smithey, Jr.; **45**CR NHPA/Anthony Bannister; **45**BL Comstock/Mike
Andrews/SGC; **46**BR Comstock/Rob Cousins/SGC; **46–7**T Planet Earth
Pictures/Hans Christian Heap; **46**CL Oxford Scientific Films/Ronald Toms;
47TR ZEFA; **48** NHPA/Anthony Bannister; **49**TR Oxford Scientific Films/
Mantis Wildlife Films; **49**BL NHPA/John Shaw; **50–1** bckgrd Science Photo
Library/Sinclair Stammers; **50**BL NHPA/Anthony Bannister; **50–1**B Oxford
Scientific Films/David Macdonald; **51**TR Oxford Scientific Films/Anthony
Bannister; **52–3** NHPA/Anthony Bannister; **52**L NHPA/Daniel Heuclin;
53BR NHPA/Nigel Dennis; **54–5**B NHPA/Peter Johnson; **55**R Oxford
Scientific Films/Owen Newman; **56–7** Comstock/Robert Frerck/Odyssey/
SGC; **56**TL Aspect Picture Library/Antoinette Jaunet; **58–9** bckgrd Andrew
Warren; **58**BR Aspect Picture Library/Tom Nebbia; **58**TL Hutchison
Library/Mick Csaky; **59**BR Impact/Colin Jones; **59**BL Magnum/© Abbas;
60–1T Comstock/Tor Eigeland/SGC; **60**TL inset Hutchison Library; **61**LC
Aspect Picture Library/Pierre Jaunet; **61**BR Magnum/© Abbas; **62–3** Robert
Harding Picture Library; **63** Comstock/John Bulmer/SGC; **64–5** ZEFA;
65BL ZEFA; **65**TR ZEFA; **65**BR Impact/Colin Jones; **66**T Comstock/Tor
Eigeland/SGC; **66–7** bckgrd Planet Earth Pictures/J.R. Bracegirdle; **67**L
Hutchison Library/Liba Taylor; **68** Frank Spooner Pictures; **69**BR
Comstock/Victor Englebert/SGC; **69**CL Magnum/© Steve McCurry; **70–71**
Magnum/© F. Mayer; **71**B Hutchison Library/Edward Parker; **72**TL Robert
Harding Picture Library; **72–3**T Impact/John Evans; **73**Bl Magnum/© Bruno
Barbey; **74**TL Comstock/Tor Eigeland/SGC; **74–5** bckgrd Shell Photo
Service; **74–5**B Comstock/Victor Englebert/SGC; **75**BR Bruce Coleman Ltd/
M.P.L. Fogden; **76–7** ZEFA; **76**BL Robert Harding Picture Library; **77**TR
Image Bank/Guido Alberto Rossi; **78–9** ZEFA; **79**TR ZEFA; **80**TL
Magnum/© Steve McCurry; **80–1** Frank Spooner Pictures; **81**BR David
Keith Jones; **82** ZEFA; **83**BL World Wildlife Fund/Photo John Newby; **83**T
NHPA/Anthony Bannister; **84**TL Geoscience Features/R. Macey; **84–5**
ZEFA; **85**BL Bruce Coleman Ltd/Dr Stephen Coyne; **86** Aspect Picture
Library; **87** Oxford Scientific Films/Ronald Toms; **88–9**B Comstock/Tor
Eigeland/SGC; **88**TR ZEFA; **89**TL Image Bank/Alvis Upitis; **90**BL Aspect
Picture Library; **90–1** EarthSat; **92**TL Magnum/© Abbas; **92–3**T Shell Photo
Service; **93**BR Shell Photo Service; **94**BL NHPA/Anthony Bannister; **94–5**
ZEFA; **95**TL inset Image Bank; **96–7** bckgrd ZEFA; **96**T Magnum/© Steve
McCurry; **97**CL Magnum/© Steve McCurry; **98–9** EarthSat; **102–3**
EarthSat; **104**CL Shell Photo Service; **104–5** Image Bank/Nick Nicholson;
105TR Panos Pictures/Penny Tweedie; **106–7** Impact/John Evans; **107**C
Aspect Picture Library; **107**B Impact/John Evans; **108–9** EarthSat; **109**TR
Magnum/© F. Scianna; **110–1** EarthSat; **112**TR Aspect Picture Library/
Antoinette Jaunet; **112–13** NHPA/Nigel Dennis; **113**BR Aspect Picture
Library; **114–5**B EarthSat; **116–7**T Comstock/Anthony Howarth/SGC;
116–7 bckgrd Shell Photo Service; **116–7**B Shell Photo Service; **118**B
EarthSat; **120–1**T Magnum/© Bruno Barbey; **120**TL Magnum/© Abbas;
121BL Frank Spooner Pictures; **122** NASA; **123**BL NASA; **124**BR Robert
Harding Picture Library; **124**TL ZEFA; **125**T Rex Features; **127**T EarthSat;
128–9T Aspect Picture Library/Peter Carmichael; **128**BL Aspect Picture
Library/Tom Nebbia; **129**TR inset ZEFA; **130**T NASA; **130**BR Comstock/
Leon Schadeberg/SGC; **132–3** EarthSat; **134–5**T Oxford Scientific Films/
Judd Cooney; **134**B Magnum/© Ernst Haas; **135**RC Comstock/Adam
Woolfit/SGC; **136–7** EarthSat; **138–9**T ZEFA; **139**BR Magnum/© Ernst
Haas; **139**TL inset ZEFA; **140–1**B EarthSat; **142**TL Image Bank/Guido
Alberto Rossi; **142–3**B Planet Earth Pictures/William M. Smithey, Jr.; **143**BR
Colorific/Michael S. Yamashita; **144–5** EarthSat; **146–7**B Impact/Alastair
Indge; **146**T Frank Spooner Pictures; **147**TR Impact/Rhonda Klevansky;
148–9 EarthSat; **150–1**T Science Photo Library/Simon Fraser; **151**B Robert
Harding Picture Library; **152–3** EarthSat; **154–5**T Comstock/Richard
Woldendorp/SGC; **155**CR Magnum/Peter Marlow; **155**BL Panos Pictures/
Penny Tweedie; **156**TL Comstock/David Austen/SGC; **156–7**B ZEFA;
158–9 Oxford Scientific Films/Kathie Atkinson; **158**TL Comstock/Richard
Woldendorp/SGC; **160–1** bckgrd Andrew Warren; **160–1** Rex Features; **161**T
World Wildlife Fund/Photo Werner Gartung/Wings; **162**L Aspect Picture
Library/Tom Nebbia; **162–3**B Comstock/Mike Andrews/SGC; **162–3** ZEFA;
164–5C ZEFA; **164**TL Magnum/© Steve McCurry; **164–5** bckgrd NHPA/
Peter Johnson; **166–7**C World Wildlife Fund/Photo John Newby; **166**TL
Magnum/© Steve McCurry; **166–7** bckgrd Andrew Warren.

Every effort has been made to trace the copyright holders and we apologise in
advance for any unintentional omissions. We would be pleased to insert the
appropriate acknowledgement in any subsequent edition of this publication.